Ecological Strategies
of Xylem Evolution

Ecological Strategies ᜂ of Xylem Evolution

by Sherwin Carlquist

University of California Press

Berkeley
Los Angeles
London

University of California Press
Berkeley and Los Angeles, California
University of California Press, Ltd.
London, England
Copyright © 1975, by
The Regents of the University of California
ISBN 0–520–02730–2
Library of Congress Catalog Card Number: 74–76382
Printed in the United States of America

For Dr. Evelyn Hooker

Contents

Preface &
Acknowledgments

A functional view of plant anatomy has been slow in developing, although many would agree on the desirability of this viewpoint. Amalgamation of facts on plant structure into a broadly oriented evolutionary synthesis is in its beginnings for several reasons.

Applicable data from plant physiology have accumulated only recently. Written at the opening of the twentieth century, Haberlandt's *Physiological Plant Anatomy* never lived up to its title—nor could it have. It tells us much more about plant anatomy than physiology; relating anatomy to function was, for plant anatomists of Haberlandt's time, a goal that could not be attained. As physiology has grown, data of use to plant anatomists have come into existence. However, these data are not aimed at the plant anatomist primarily, and therefore often are awkward material for answering questions about structure. Moreover, plant anatomists are not familiar with physiological literature, and there has been little dialogue between these fields.

Structural engineering has much to offer the plant anatomist. I have responded to this challenge in only a crude way, largely by way of applied structural engineering, and specifically with analogies from architecture. Ideally, construction of mathematical models of morphological and anatomical structures would be desirable. These structures represent approaches to what an engineer would term optimal structure models. Mathematical development of these models would require much more than a single book. By consulting a structural engineer, I discovered how extremely complex this task would be. Such an effort would be an interesting enterprise. Had I attempted it here, the results would be largely incomprehensible to plant anatomists. Those who are interested in this approach should consult a recent paper

(Banks, 1973) which outlines the problems and rewards of the engineering viewpoint.

With limitations such as the above, one may say that the present essay is at worst either a restatement of the obvious, or a series of speculative excursions for which no solid evidence exists. The former, if true, would be harmless; the latter would underline how little we know and indicate areas for resolution. At best, this book may permit hitherto unappreciated correlations to emerge.

These syntheses are essential precisely because of the manner in which fields that ought to be in constant contact have, instead, progressively grown apart. This phenomenon is a byproduct of the perpetual diversification of science. For example, the study of floral venation has become a field with its own methodology, omitting correlations with floral nutrition, pollen presentation mechanisms, pollination methods, fruit type and size, and dispersal mechanisms. Complicated hypotheses for which there is no basis in fact and which are functionally unlikely have been advanced for processes of evolution of flowers.

One phase of diversification in plant anatomy has featured comparative studies of wood anatomy. Xylem of non-woody plants has also been studied. Comparative wood anatomy has, in large measure, emanated from forestry, and has had identification of timbers as a frequent goal. This focus is entirely understandable and much useful data have been produced. In the process of accumulation of these data, however, the functional nature of xylem has been left relatively unexplored.

An evolutionary synthesis of plant anatomy requires an understanding of form and function and ecology as projected through time via phylogeny; in short, a preposterously broad knowledge of numerous fields. To be adequate, such a synthesis would require original data; for example, data on ecological tolerances are available for only a few species of a plant. Ecological observations offered here are, of necessity, crude generalizations. However, the correlations of these vague data with facts from plant anatomy are compelling. With respect to comparative studies, one must take into account a large number of genera. Many genera are mentioned in this book; families to which these belong are indicated in the Index.

Constructing an evolutionary synthesis of structure, function and ecology is a task as difficult as it is valuable. No single person can accomplish it; even a team of individuals is unlikely to provide a total portrait of the functional significance of xylem. Those who engage in this enterprise will find that the most diverse kinds of data are applicable. I have cited such phenomena as mycorrhizal associations, plugged stomata, soil porosity, and leaf size. The diversity of potential correlations suggests that I have neglected many types of data and undoubtedly a number of papers and books that might have furthered this effort.

Several individuals have aided me in the production of this book. Dr. William L. Stern has kindly read a preliminary manuscript and offered many helpful suggestions. Mr. David Wheat spent two years preparing some of the slides and data utilized in this book. Dr. Edwin A. Phillips offered valuable aid in literature on plant physiology. Dr. Harry E. Williams provided insights in the fields of structural engineering and architecture. Mr. Larry De Buhr and Dr. Gary Wallace made preparations of gymnosperm woods. Others who offered useful comments and encouragement include: Dr. Edward S. Ayensu, Dr. Vernon Cheadle, Dr. David F. Cutler, Dr. Leo Hickey, Dr. Harlan Lewis, Dr. C. R. Metcalfe, Dr. Robert Ornduff, and Dr. Margaret L. Stant. Studies basic to this book have been aided by grants from the National Science Foundation (particularly NSFG-23396 and extension thereof; and earlier grants GB-4977x and GB-14092). Field work in Malaya and South Africa that contributed to materials in this book was sponsored by a Fellowship from the John Simon Guggenheim Memorial Foundation, as well as a recent National Science Foundation Grant, GB-38901.

Introduction

Various good contributions have been made to development of principles of xylem evolution. These studies have expressed results in terms of "primitive" versus "specialized" modes of structure. What we must now understand is *why* xylem of particular species exhibits features designated as primitive or specialized. In other words, what factors induce progressions toward advanced xylem characteristics? Why should species with "primitive" xylem configurations persist at all, and how have they done so? Is there any degree of reversibility in xylem evolution? If so, what forms does it take? Statistical correlation can be shown among certain features (e.g., scalariform lateral wall-pitting correlated with scalariform perforation plates on vessels of dicotyledons). However, plants do not evolve because of statistical correlations, but because a particular feature or constellation of features is adaptive in a given situation. If anything, species that show deviations from statistical correlations (and there are always many such species) may tell us more about the adaptive value of particular characteristics than ɔ species that fit the correlation.

The keys to xylem evolution that have not been appreciated hitherto are primarily in the following areas: adaptation to degree of moisture availability and transpiration rate; fluctuation in moisture availability; and requirements for mechanical strength. Noting my (1966a) review of ecological factors related to wood evolution in Asteraceae (Compositae), Bailey (1966) stated, "It should be emphasized here, that in future investigations of anatomical differences in plants of divergent habits of growth more attention should be devoted to ecological and physiological influences in the habitats in which plants normally grow." This admirable statement suggests comparison of species within a genus, or other surveys within a diversified group. That is definitely necessary, and is basic to my earlier studies and to many of the examples cited in this book. However, I have discovered that the widest possible comparisons are required, for only then can one be assured of satisfying explanations. Each group of vascular plants presents evidence concerning the way ecological factors influence xylem evolution, and therefore I have been unable to maintain a narrow approach.

In relating water availability and mechanical strength to xylem anatomy, I must emphasize that unfortunately no single measure can be used, nor can we relate several factors by means of a multiple regression equation. Conceivably, for a given plant, a multiple regression equation could be devised to explain the xylem pattern of a particular portion of a plant in a particular stage of development. Such designs, if desirable, ought to follow elucidation of the various major patterns. For example, what has influenced the xylem conformations of *Allium* is quite different from what has influenced the wood pattern of *Sequoia*, although these two genera might be found growing together. Where they do occur together, they utilize the environment in quite different ways.

The correlations that hold true for one group of vascular plants may not apply to another group. The major groups of vascular plants differ in whether secondary xylem is produced (gymnosperms, dicotyledons) or not (living pteridophytes except *Botrychium* and *Helminthostachys*; monocotyledons—monocotyledons in which secondary activity is present do not have true secondary xylem). Gymnosperms (except Gnetales) and vesselless dicotyledons have a single multipurpose tracheary element, the tracheid. The remainder of dicotyledons have division of labor between vessel elements and imperforate tracheary elements (tracheids, fiber-tracheids, or libriform fibers) with few exceptions (e.g., *Crassula argentea*, plate 13-C, D). This familiar situation has interesting implications. Ferns, other pteridophytes, and monocotyledons have independent conducting tissue and mechanical tissue; however, one must stress that fibrovascular bundles in these groups are definitely related when bundle-sheath fibers and mechanical tissues (extraxylary fibers) are both derived from given procambial strands. However, this is quite different from derivation of both vessel elements and libriform fibers from a single fusiform cambial initial in dicotyledons. Thus, tracheary tissue in ferns and monocotyledons can be studied directly with respect to the relationship of conducting tissue to water availability. Tracheids of gymnosperms and vesselless dicotyledons demonstrate both conductive and mechanical capabilities, and analysis of their wood structure must discern which features are governed by mechanical considerations, which by conductive considerations. In vessel-bearing dicotyledons and Gnetales, mechanical elements (imperforate elements) and cells of conductive efficiency (vessel elements) are both derived from fusiform cambial initials, so that lengths are not totally independent. However, there can be wide divergence, in any given wood, in length of imperforate elements and vessel elements so that ecological factors can, theoretically, operate differently on each cell type. This independence is more operative in the case of diameters of vessels and wall characteristics of imperforate elements. In dicotyledons, however, there are complicating features such as ray type and histology, and more intricate adaptations are possible. Nevertheless, the major groups of vascular plants show ecological significance when analyzed in terms of families and genera within them and when analyzed with relationship to each other.

Reluctantly, I have presented a discussion of stelar types in vascular plants. Although I feel information requisite for an adequate revision of the significance of stelar types is not yet

at hand, I find that analysis of cells of conducting tissue requires a broader context; specifically, the disposition of that
conducting tissue in roots, stems, and petioles.

Some anatomical data are presented here in quantitative
form. However, even these figures cannot be precise, as any-
one who understands xylem anatomy can well comprehend.
Limitations to precision in measurements are stressed ex-
plicitly and implicitly by the data of Bailey and Tupper
(1918), Bailey and Faull (1934), Spurr and Hyvärinen
(1954), Stern and Greene (1958), Dinwoodie (1961), Rumball (1963), De Zeeuw (1965), Bannan (1965, 1967, 1968),
and Sastrapadja and Lamoureux (1969). Even within a single
growth ring, variation in element length can be considerable
(Swamy, Parameswaran, and Govindarajalu, 1960). However,
my interest in these various studies does not lie in the difficulties of presenting measures for anatomical features of a
species and for portions of a plant, but rather in what these
variations mean in terms of adaptations by the plant.

HISTORICAL REVIEW

Variation in Quantitative Characteristics

Variation within an individual plant.—Contributions applicable to the present hypotheses are mostly from workers interested in forestry and, to a lesser extent, physiology, ecology,
and anatomy. While many of these contributions are good,
they are non-additive and new data are required. The patterns
that have been reported demand a coherent explanation.

Sanio (1872) discovered that in *Pinus sylvestris* the following patterns occur:

(1) Tracheids of secondary xylem are shorter at the inside
of the trunk or branches, and increase in length through a
number of annual rings toward the outside. Often, a plateau
in length is obtained in older stems.

(2) The length of tracheids increases from the base of the

plant toward the top. However, tracheid length reaches a maximum at a certain height, and shorter tracheids characterize upper branches.

These, as well as other findings, were termed "Sanio's Laws" by Bailey and Shepard (1915). Numerous other studies, summarized by Spurr and Hyvärinen (1954), Dinwoodie (1961), and De Zeeuw (1965) tend to show much the same thing. Decrease in tracheid length in outermost annual rings of conifers was reported by Bailey and Tupper (1918) and Bannan (1967) and attributed to senescence. Bailey and Faull (1934) demonstrated Sanio's Laws in *Sequoia sempervirens*. In addition, they found that tracheids tend to be longer in roots (fig. 7) than stems. Patterns of vessel-element length and fiber length within a palm stem were presented in tabular form by Swamy and Govindarajalu (1961) and have been graphed here as figure 10. Similar data for *Sabal palmetto* were presented by Tomlinson and Zimmermann (1967). In general, the variations noted in Sanio's Laws have been confirmed for a number of conifers and dicotyledons (Spurr and Hyvärinen, 1954), although inconsistencies have been noted.

Wider growth rings tend to have shorter tracheids in mature trees in conifers studied by Bannan (1965). Within individual growth rings, greater length tends to characterize latewood tracheids in conifers, and latewood libriform fibers in dicotyledons, although these patterns are not without exception (see Spurr and Hyvärinen), a fact visible from the curves for variation in growth rings presented by Swamy, Parameswaran, and Govindarajalu (1960). One must remember in viewing these inconsistencies that accurate determination of element length within a single growth ring may be extremely difficult and unreliable, and one is tempted to accept with caution the generalization that imperforate elements are longer in latewood in dicotyledons. One might also mention that identification of what portion of a growth ring is earlywood, what portion latewood can be made best in transections; but transections cannot be used for determination of element length. Chemical changes, which are related to mechanical properties, occur within growth rings. Lignin is more abundant in earlywood, cellulose is more abundant in latewood for both hardwoods and conifers (Ritter and Fleck, 1926).

The phenomenon of ring porosity is too well known to require further description. However, it is significant in interpretation of tracheary element diameter in woody plants at large. We do not have much data on variation in diameter of tracheids within growth rings of conifers, although Bannan's (1965) work gives good data. The interest in diameter of elements in woody plants is based on the fact that variations in diameter are independent of cambial initial length. Lengths of mature elements are related to lengths of the fusiform cambial initials from which they were derived, and may show elongation after derivation by means of intrusive growth, as in the libriform fibers of dicotyledons. However, diameters of vessel elements in dicotyledons are capable of almost unlimited widening after derivation of elements from cambial initials.

The studies of Cheadle (1942, 1943a, 1943b, 1944, 1955, 1963, 1968, 1969, 1970; Cheadle and Kokasai, 1971, 1972) on tracheary tissue within monocotyledons were directed primarily at demonstrating organographic distribution of vessel elements in a large sampling of species. In the present context, we can note that he found more specialized vessel elements to be shorter, in general. Therefore, because of the specialization sequence he presents, from roots into stems, inflorescence axes, and leaves, we could generalize that shorter vessel elements would tend to occur in roots, progressively longer elements in stems, inflorescence axes, and leaves. However, the morphological component of this specialization sequence proves to be more significant than the lengths of the tracheary elements involved.

In ferns, White (1963a) found various patterns, but generalized that tracheids tend to be longest in petioles, intermediate in roots, and shortest in rhizomes. White's data are subjected to further analysis in chapter 2. Where plants of different ages could be compared (*Osmunda cinnamomea*), White found appreciably shorter tracheids in older plants.

Age-on-tracheary-element length curves have been presented for woody species (Bailey and Tupper, 1918; Bailey and Faull, 1934; Carlquist, 1962; Anderson, 1972; Baas, 1973). Similar data, not presented in the form of graphs, have been offered by Cumbie (1963, 1967a, 1967b, 1969). These curves have aided in the development of concepts of growth curves typical of conifers and woody dicotyledons, discussed below. By contrast, curves characteristic of dicotyledons that show juvenilism, or paedomorphosis, have been identified (Carlquist, 1962).

Sastrapadja and Lamoureux (1969) report statistically significant differences in cell size in woods from the base to the top and from the center to the outside of the trunk in trees of *Metrosideros collina* (= the *M. polymorpha* complex) in the Hawaiian Islands.

From base to upper portions, they found that number of vessels per sq. mm. increases; vessel-element length and fiber-tracheid length decrease; vessel-wall thickness decreases; fiber-tracheid wall thickness increases; and ray height increases.

From center to outside in a trunk, they found that number of vessels per sq. mm. increases; vessel-element length and fiber-tracheid length decrease; and fiber-tracheid wall thickness increases.

Ontogenetic changes in tracheary-element length and diameter in vascular plants during the sequence from protoxylem to metaxylem are familiar to plant anatomists. The stylized graphs of figures 15 and 16 represent these sequences.

Variation within a species.—In particular conifer species, Bannan (1965) has found differences among populations in tracheid length in stems of comparable age. In general, one could summarize his results by saying that plants in optimal habitats have longer tracheids. Longer tracheids were reported

for fast-growing individuals of *Pinus densiflora* by Hata
(1949). Dinwoodie (1963) showed heritable differences in
tracheid length in *Picea sitchensis*: tracheid length was progressively shorter in individuals from progressively more north-
erly latitudes. Richardson (1964) has shown the experimental
optima for night temperature, day length, and so on, that
produce long tracheids in *Picea sitchensis*, and has demonstrated that tracheids are shorter in plants grown under suboptimal conditions. Swampy soils are less favorable than
sandy soils for *Thuja occidentalis*, and thus shorter tracheids
in the swamp populations are understandable (Bannan, 1941,
1942). Various workers have uniformly reported shorter tracheids in stunted or injured conifers (e.g., Bailey and Tupper,
1918).

In dicotyledons, longer libriform fibers were observed in
individuals of *Juglans californica* in more mesic sites (Mell,
1910). Longer imperforate elements were reported in more
southerly populations of *Liquidambar styraciflua* (Winstead,
1972). Differences in cell wall to lumen ratio and in cell-wall
thickness have been reported in *Shorea* (Aung, 1962). In
individuals, cell-wall thickness tends to increase toward the
outside of *Shorea* individuals. Sastrapadja and Lamoureux
(1969) report marked variations in cell dimensions among
individuals in the Hawaiian *Metrosideros* populations they
compared; few of these variations showed statistical correlation, however. Individuals from montane bogs did prove to
have shorter, narrower vessel elements. Individuals with small
pits in lateral walls of vessels were mostly from areas of low
rainfall, those with large pits from areas of rather high rainfall.

Differences among taxa and regions and their significance.—
The first major comparative investigation on sizes of xylem
cells was that of Bailey and Tupper (1918), who surveyed all
major groups of vascular plants with respect to tracheary-element length. The lack of quantitative data on tracheary elements both before and after this paper's appearance is sur-

prising to me. There are many works that give anatomical descriptions of woods. Some of these offer quantitative data, but usually not of the sort that would be pertinent to the present study. For example, Phillips (1941) and Greguss (1955) offer detailed descriptions of gymnosperm woods, but no measures of tracheid length or diameter. Some papers offer ranges of lengths for tracheary elements, but no averages. Averages for various quantitative measures are far more important than the extremes because averages reflect more accurately than any other quantitative expression the actual functional capability of any given xylem feature. More significantly, averages can be compared for portions of a plant, different populations, or different species. The presentation of qualitative and quantitative data on wood anatomy in tabular form provides ease of comparison, as well as a greater degree of precision. Regardless of the aim of the study, workers in wood anatomy should be encouraged to present their results in this fashion, and I note with satisfaction that this practice is on the increase.

When dealing with differences between species, one inevitably becomes involved with differences between geographical regions, for species differ in habitat preferences. For this reason, both types of differences are discussed here.

Tabular comparisons in my work on Asteraceae (1966a, and the papers which preceded it), Goodeniaceae (1969a), Campanulaceae subfamily Lobelioideae (1969b), Echium (1970a), Euphorbia (1970b), and Brassicaceae (1971) have led to the present concern with ecological and physiological factors. In those papers, I found the following features correlated with increased xeromorphy: narrower vessel elements, shorter vessel elements, more numerous vessels per group, shorter imperforate elements, and shorter rays.

Only a few papers have offered ecological correlations in past decades. Starr (1912) reported fewer vessels per sq. mm. in mesophytic individuals of Alnus incana compared to those from xeric sites. Kanehira (1921a) called attention to the problems of relating wood anatomy to climate. Webber

(1936) showed notably short lengths of vessel elements in desert and chaparral shrubs of southern California, although she did not offer data on mesomorphic counterparts. Cumbie (1960), Cumbie and Mertz (1962), Ayensu and Stern (1964), and Walsh (1974) have related xylem anatomy to habit of growth. Tabata (1964) related habit and habitat data to vessel morphology, particularly number of bars per perforation plate and number of vessels per unit area of transection (but not tracheary element dimensions) in Japanese *Betula* species. Baas (1973) has rightly questioned the validity of Tabata's conclusions, since Tabata's materials were inadequate. Gibson (1973) has produced an admirable study of woods in the cereoid cacti in which correlations between habit and wood anatomy are stressed.

Versteegh (1968) has shown that in the woody montane flora of Indonesia, there are Lauraceae, Anacardiaceae, and Casuarinaceae with scalariform perforation plates. In the lowlands of Indonesia, species from these families have simple perforation plates. In addition to these quantifiable differences, Versteegh found that in Apocynaceae and Myrtaceae there are montane species with tracheids, whereas libriform fibers occur in the wood of the lowland species of these families.

Novruzova (1968), in a survey of wood anatomy of Azerbaijan trees and shrubs noted shorter, narrower vessels with simple perforation plates in plants of drier areas; whereas longer, wider vessels and a higher proportion of species with scalariform perforation plates could be found in woods of mesic areas. Borders on pits of imperforate elements were reported to be smaller or absent in imperforate elements of species from drier habitats compared with those from wet sites. Where a species with scalariform perforation plates grew in varied sites, fewer bars per perforation plate were observed in the individuals of drier habitats. At Baas (1973) notes, there is close agreement between the conclusions offered by Novruzova (1968) and the observations in my various papers

(1966*a*, 1969*a*, 1969*b*, 1970*a*, 1970*b*, 1971), despite the fact that Novruzova worked with a floristic region and I worked with particular groups of dicotyledons.

Baas (1973) has stressed the relationship between latitude and various wood characteristics in his analysis of quantitative features of species of *Ilex*. With increasing latitude, he found decrease in vessel-element length, decrease in number of bars per perforation plate, a decrease in vessel diameter, an increase in pit size and border width on tracheids, and increase in number of vessels per sq. mm. Although Baas discriminates in his discussion between lowland tropics and upland tropics, his graphed data on *Ilex* are presented only in terms in latitude. However, he does incorporate altitude in his data on *Prunus*, in which vessel-element length decreases with increase in latitude. Greater vessel-element lengths near the tropics and shorter ones in temperate zones have been shown by Schweitzer (1971) in *Celtis* and Dadswell and Ingle (1954) in *Nothofagus*.

The above studies are commendable, although they do not offer the sort of data base one would wish for comparisons of structure with ecology and physiology. The shortage is partly in quantitative data, but more severely in availability of ecological and physiological data for the species studied with respect to anatomy.

Morphological Aspects of Xylem Evolution

The study by Bailey and Tupper (1918) provided a tool—tracheary-element length—which is quantitative yet can be applied to analysis of morphological features. The most important of these analyses were performed not by Bailey himself, but by his students. Bailey and Tupper concluded that "tracheary elements of vascular cryptogams tended to be very long, whereas those which occur in the dicotyledons—with the notable exception of the vesselless Trochodendraceae and [Winteraceae]—are comparatively short. The gymnosperms appear to occupy an intermediate position between these ex-

tremes; the Cordaitales [Cycadeoideales], and Cycadales
resembling the vascular cryptogams, and the Gnetales—sup-
posed gymnosperms with vessels simulating the angiosperms."
Subsidiary conclusions of Bailey and Tupper have been stated
more concisely by Bailey (1953): "Two means of accelerat-
ing conduction have been adopted: (1) increasing the length
of large, thin-walled profusely pitted tracheids; and (2) developing vessels which lead to a shortening of tracheary cells."

Origin of vessels in dicotyledons.—The key element of tracheary-element length was used by Frost to demonstrate
origin of vessels in dicotyledons (1930*a*), evolution of the
vessel end wall (1930*b*) and evolution of the lateral wall of
the vessel element (1931). Frost's method of correlation and
his conclusions are simple and require little modification. On
the origin of the vessel element, Frost (1930*a*) concluded that
the woody vesselless dicotyledons (e.g., Winteraceae) are
primitively vesselless. He believed that primitive vessel elements resemble tracheids in great length, angularity of transectional outline, smallness of diameter, absence of a distinct
end wall, and presence of thin, evenly thickened end walls.
Frost concludes that vessels were derived from a wood consisting of scalariformly pitted tracheids, a conclusion shared by
Bailey (1944*a*). Later in this book, I have offered a modification to that particular conclusion. Also, one may note that
vessels angular in transection occur in specialized woods
(Carlquist, 1962; Sastrapadja and Lamoureux, 1969). Narrowness of vessels by itself bears no relationship to phylogenetic
level of specialization.

Specialization of vessel elements in dicotyledons.—With respect to specialization of the perforation plate, Frost (1930*b*)
concluded that there is a progression from scalariform oblique
to simple transverse. He reached this and other conclusions by
selecting groupings along this gradient, and determining the
average vessel-element length for each grouping of species.

The progression from scalariform oblique to simple transverse proved to be paralleled by decreasing length of vessel elements. Frost also concluded that presence of borders on apertures of scalariform perforation plates is more primitive than absence of borders. He claimed that widening of the apertures of the scalariform perforation plate was a trend of specialization; this could have been predicted, since widening of the apertures inevitably involves diminution of number of bars as a generalization.

Frost concluded that lateral-wall pitting progresses more rapidly than simplification of the end wall, because of the 51 species with scalariform plates, 22 do not have scalariform lateral-wall pitting, but opposite or alternate pitting instead. However, there are exceptions to this trend. For example, figured here are *Michelia fuscata* (plate 8-F), which has prominent scalariform lateral-wall pitting but only about four bars per plate; and *Ceriops tagel* (plate 14-D), which has four to six bars per perforation plate, and only scalariform lateral-wall pitting. These two species are typical of their respective families, Magnoliaceae and Rhizophoraceae, in these respects. There are even dicotyledons with simple perforation plates and scalariform lateral-wall pitting (Carlquist, 1962). These exceptions do not show up in the rather coarsely designed correlations of Frost. His data do not show deviations, only averages. For example, his figures show that species with "oblique porous" (porous = simple) perforation plates average 0.69 mm. in vessel-element length, and "transverse porous" species average 0.41 mm. *Corokia cotoneaster* has an average vessel-element length of 0.56 mm. (Patel, 1973), and thus ought to have simple perforation plates between oblique and transverse in orientation. In fact, it has scalariform perforation plates with about ten bars. Also, we might notice that Frost gives no list of the species he studied, nor any reasons either why the progression he demonstrates has taken place or why the primitive types still persist—or what percentage of dicotyledons the primitive types constitute.

Frost (1931) designated four groups of species on the basis of lateral-wall pitting of vessels, and found them to be correlated with progressively shorter average vessel-element length: scalariform (1.13 mm.); transitional (1.07 mm.); opposite (0.79 mm.) and alternate (0.46 mm.). He found both intervascular and vessel–ray pitting to follow the same trend.

Frost (1931) also claimed progression of fully bordered pit pairs to half bordered pit pairs to simple pit pairs between vessels and parenchyma. This is perhaps surprising in view of the stereotyped notion one sees in textbooks that half bordered pit pairs characterize vessel–parenchyma pitting. Frost hypothesizes the fully bordered pit pair as primitive largely by default. However, gymnosperms typically have half bordered pit pairs between tracheids and parenchyma. An exception to this occurs in *Podocarpus macrophylla* in the case of particularly thick-walled ray cells that are related to tracheids by fully bordered pit pairs (Greguss, 1955). A number of conifers have ray tracheids that are related to true tracheids by fully bordered pit pairs. I have observed fully bordered pit pairs among the ray cells of all genera of vesselless dicotyledons (plate 7-D). Wall thickness of ray cells in these genera is as great as, or greater than, that of tracheids. In one genus (*Amborella*) I was able to observe presence of fully bordered pit pairs between ray cells and tracheids. On the other hand, fully bordered pit pairs may be seen in ray cells of specialized woods, such as *Metrosideros* (Sastrapadja and Lamoureux, 1969).

Frost (1931) also concluded that spirals on secondary xylem vessels, sometimes termed "tertiary helical thickenings," are not primitive in dicotyledons. These thickenings do occur in woods with a wide range of degree of specialization. Kanehira (1921*a*, 1921*b*, 1924) found that a smaller percentage of tropical woods have spirals, a relatively high percentage of temperate woods. Schweitzer (1971) also reported this for *Celtis*. Baas (1973) finds increase in spirals with increasing latitude in *Ilex*, and increase in spirals with both increasing altitude and latitude in *Prunus*.

Ring porosity was claimed by Gilbert (1940) to be specialized compared with diffuse porosity. This conclusion can be supported easily enough, with the caution that it is a specialization that occurs in woods of various levels of phylogenetic advancement based on vessel characteristics. However, Gilbert's contention that ring porosity evolved early in the history of angiosperms, and his claim that the phenomenon is a north-temperate phenomenon can be easily disproved. For example, there are numerous ring-porous species of woody Asteraceae, some of which occur in the Southern Hemisphere. Although Southern Hemisphere woods are not as well known as those of the Northern Hemisphere, other Southern Hemisphere species that are ring porous can easily be cited.

Parenchyma evolution in dicotyledons.—A logical extension of the length and morphology of the vessel element to problems in wood evolution was the application by Kribs to wood rays (1935) and axial parenchyma (1937). Kribs used both vessel-element length and end-wall angle, as shown in tables 1, 2, and 3, to determine the evolutionary interrelationships among the ray types he designated, which have been drawn in figure 1.

The progression from heterocellular (heterogeneous) to homocellular (homogeneous) rays is seen clearly in tables 2 and 3. The reader unfamiliar with ray histology should note that square cells are grouped with upright cells for the purpose of distinguishing between these two types. Identification of these types can be made reliably only in radial sections, as shown by the diagrams to the right of rays in figure 1. Kribs's data show both progression to a homocellular condition and progression to presence of either uniseriate or multiseriate rays exclusively. The small number of genera with uniseriate rays exclusively (Heterogeneous III and Homogeneous III), as given in table 3, suggests that absence of multiseriate rays is not as adaptive as it is in conifers, where uniseriate rays are present exclusively except where secretory canals occur.

TABLE 1
Percentage of Ray Types in Each Vessel Type

Type of Vessel Element	Number of Genera	Ray Types					
		Multiseriate and Uniseriate				Uniseriate Only	
		Heterogeneous		Homogeneous		Hetero-geneous Type III	Homo-geneous Type III
		Type I	Type II	Type I	Type II		
Scalariform I............	63	79.36	15.87	...	...	4.77	...
Scalariform II............	32	53.12	21.87	12.50	3.13	9.38	...
Scalariform-porous......	67	47.77	35.81	5.98	4.48	2.98	2.98
Porous-oblique..........	220	44.09	43.18	5.45	3.56	2.36	1.36
Porous-oblique and transverse............	148	9.45	33.11	34.45	9.49	2.02	11.48
Porous-transverse........	220	...	19.09	27.28	44.10	1.81	7.72

SOURCE: Kribs, 1935.

TABLE 2
Total Percentage of Ray Types in Each Vessel Type

Type of Vessel Element	Percentage Rays	
	Heterogeneous	Homogeneous
Scalariform I......................	100.00	...
Scalariform II.....................	84.37	15.63
Scalariform-porous.................	86.56	13.44
Porous-oblique.....................	89.63	10.37
Porous-oblique and transverse.......	44.58	55.42
Porous-transverse..................	20.90	79.10

SOURCE: Kribs, 1935.

Kribs's typology was designed to take account of the fact that the two types of progressions (to homocellular rays; to multiseriate or uniseriate rays exclusively) take place simultaneously. Table 1 justifies the belief in these progressions, but forces one to conclude that the ray types are not precisely correlated with vessel types. In other words, table 1 shows that "primitive" rays may be found in conjunction with specialized

TABLE 3
Average Length of Vessel Element for Each Ray Type

Type of Ray	Number of Genera	Average Vessel-Element Length (mm.)
Heterogeneous Type I................	210	0.81
Heterogeneous Type III.............	18	0.64
Heterogeneous Type II..............	227	0.58
Homogeneous Type I.................	131	0.52
Homogeneous Type III..............	41	0.38
Homogeneous Type II..............	123	0.35

SOURCE: Kribs, 1935.

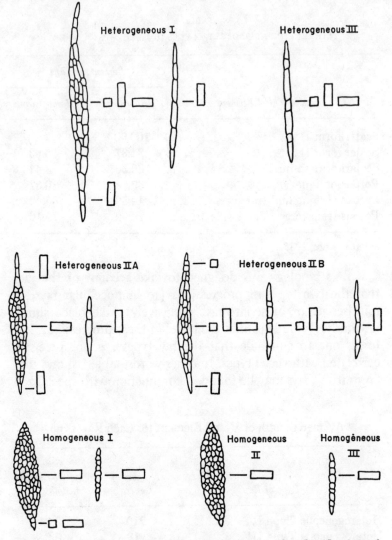

Figure 1. Ray types in dicotyledons according to the classification of Kribs (1935). For each type, tangential sections of rays are shown. To the right of each are stylized drawings of erect, square, and procumbent cells as they would be seen in radial section. These are shown for wings of rays, where present, as well as for the central portions of rays. In the Heterogeneous II B type, either of the two types of uniseriate rays may occur.

vessels in some ray types in particular species; conversely, "specialized" ray types may be found in conjunction with primitive vessels in certain species. Kribs does not provide us with the list of species he utilized, nor their ray types, respectively. Therefore, his data cannot be reanalyzed. Also, Kribs did not include any species in his study in which ray cells are square and erect exclusively, or erect exclusively. Rays of these types can be found in such families as Asteraceae (Carlquist, 1966a) and Campanulaceae, (Carlquist, 1969b) rather frequently. The ontogenetic significance of such rays has been discussed elsewhere (Carlquist, 1962, 1966a).

Barghoorn's studies on rays (1940, 1941a) gave an important ontogenetic dimension to our knowledge of rays. He demonstrated, for example, that in a given species, rays can lose heterogeneity (heterocellularity) during stem growth. Barghoorn (1941b) also explored the phenomenon of raylessness.

Braun (1970) has contributed a complicated volume on wood histology. With regard to rays, he takes advantage of a distinction, little noted previously, between "contact cells" and "isolation cells." Contact cells are ray cells in which pits are prominent on walls facing vessels. Isolation cells are ray cells with a few, small pits where they face vessels. In addition to these two types, Braun recognizes inner cells of multiseriate rays as a third category. By combining these three types of ray cells with other xylem features, Braun constructs an intricate series of structure types and subtypes. Although these constructs are the main thrust of Braun's book, he does imply physiological significance.

Axial parenchyma evolution in dicotyledons was studied by Kribs (1937), who presented correlations between classes of vessel end walls and axial-parenchyma types (table 4) and between parenchyma types and vessel-element lengths (table 5). In table 4, we see close correlation for some parenchyma types, little for others. For example, "parenchyma absent" shows little correlation. Lower percentages for that feature among woods with specialized vessel types could well be an artifact of species studied by Kribs. For example, axial parenchyma is absent or very scarce in Cistaceae, Connaraceae, Gesneriaceae, Polygonaceae, Solanaceae, Verbenaceae, and Vitaceae. I suspect these families are under-represented in Kribs's survey materials. The type of parenchyma termed "vasicentric scanty" by Kribs is omnipresent in the family Asteraceae, in which "porous-transverse" end walls are frequent, yet table 4 shows no percentage of species with both porous-transverse end walls and vasicentric scanty parenchyma.

One can draw a general conclusion from Kribs's data that diffuse axial parenchyma is primitive, and that the various types of paratracheal ("vasicentric") parenchyma are specialized. Kribs, however, presented a detailed phylogenetic tree of axial parenchyma types. His phylogeny probably represents only a portion of the pathways possible, and additional possibilities have been indicated by Bailey and Howard (1941). Bailey (1957) cites the possibility that bands of parenchyma in certain legume woods have arisen not from phylogenetic regrouping of parenchyma cells, but rather from maturation of derivatives of the cambium into long-lived libriform parenchyma cells ("nucleated fibers" of some authors) rather than short-lived libriform fibers. This phenomenon occurs also in certain Asteraceae (Carlquist, 1958), where degrees of differentiation among the types of libriform elements in closely related species and genera permit one to term this phenomenon fiber dimorphism (Carlquist, 1961a).

The system of Braun (1970, fig. 88) for organization of axial parenchyma incorporates five stages, through which three series are envisaged as evolving independently. These "stages" and "series" are probably not intended to be literally phylogenetic, only generally so. Particular genera are cited for each of these types. Braun's stages do, however, agree with the general conclusions to be derived from Kribs's work.

Further considerations on ray and axial parenchyma evolution will be found in chapter 11.

TABLE 4

Percentage of Parenchyma Types in Each Vessel Type

Type of Vessel Element	Number of Genera	Parenchyma Types							
		Diffuse	Diffuse Aggregate	Vasicentric Scanty	Metatracheal Narrow	Metatracheal Wide	Vasicentric Abundant	Parenchyma Absent	Terminal
Scalariform I........	63	69.84	19.04					11.12	
Scalariform II.......	32	59.37	15.62					12.50	12.51
Scalariform-porous..	68	13.23	41.18	20.59	8.82			11.76	4.42
Porous-oblique......	202	11.88	32.65	15.78	8.41		11.96	12.87	6.45
Porous-oblique and transverse.........	192	4.16	14.06	7.81	17.18	10.42	31.28	4.16	10.93
Porous-transverse...	225		2.69		6.66	9.33	70.22	1.77	9.33

SOURCE: Kribs, 1937.

TABLE 5

Average Length of Vessel Element for Each Parenchyma Type

Type of Parenchyma	Number of Genera	Average Vessel-Element Length (mm.)
Diffuse..........................	104	.92
Absent...........................	57	.78
Diffuse-aggregate...................	144	.65
Vasicentric scanty...................	61	.60
Metatracheal narrow...............	71	.51
Terminal..........................	62	.44
Metatracheal wide.................	41	.42
Vasicentric abundant...............	242	.31

SOURCE: Kribs, 1937.

Xylem of monocotyledons.—The application of vessel-element length and morphology to study of evolution in dicotyledons naturally led to use of these findings in monocotyledons. Because maceration techniques are less easy in monocotyledons than in dicotyledons and because roots, stems, leaves, and inflorescence axes had to be studied independently, monocotyledons proved more difficult to attack. However, Cheadle presented a successful survey in a series of four papers (1942, 1943a, 1943b, 1944). In these papers convincing phylogenetic conclusions, which still appear thoroughly valid, were presented. Cheadle and others have expanded the data base available in papers cited later. Cheadle found the presence of tracheids and vessel elements in various portions of plants a prime tool for study of phylogenetic sequences. If vessels were present, the nature of the perforation plate (degrees from scalariform to simple) proved of prime importance. Length of tracheids or vessel elements was of quite subsidiary importance, for reasons discussed later. Cheadle showed that, phylogenetically, vessels originated first in roots of monocotyledons and progressed into stems, inflorescence axes, and leaves, in that order. The order for morphological specialization of vessel elements, if present above roots, follows the same sequence. However, Cheadle (1944) has taken note of the fact that in some organs, vessel specialization may be more accelerated, in others more retarded, rather than in a perfect gradation.

Primary xylem of dicotyledons.—Bailey (1944*b*) hypothesized that vessels originated in secondary xylem of dicotyledons, and were introduced simultaneously into roots and stems. As a corollary, he concluded that vessels progressed from secondary into primary xylem of dicotyledons. Evidence for this was provided by the studies of Bierhorst (1960) and Bierhorst and Zamora (1965). Bierhorst showed that in dicotyledons with relatively primitive secondary xylem but vessels in secondary xylem, tracheids only are formed in primary xylem, at least the earlier portions. At the other extreme, only taxonomic groups with highly specialized secondary xylem proved to have vessels in the primary xylem also, and only a portion of these groups had vessels with simple perforation plates in primary xylem. In addition, however, Bierhorst's studies reveal that a wide variety of types may be found in end-wall and lateral-wall morphology. This range is wider, more variable, and more complex than might have been expected, and is a valuable feature of Bierhorst's contribution. Some of the variations appear to have no precise phylogenetic or functional correlates. The logical conclusion to be drawn is that a variety of peculiar types are functionally quite effective. The general outlines of his conclusions, however, are not changed by taking into account these variant types. In the two papers cited, primary xylem variations of groups other than dicotyledons are considered.

In 1962, I proposed a "theory of paedomorphosis in dicotyledonous woods." This idea was formulated to explain how types of vessel-element length and lateral-wall pitting characteristic of primary xylem occur in secondary xylem and only

gradually yield, if at all, to a "mature" pattern. Patterns re-
ferable to paedomorphosis can be found in stem succulents,
rosette trees, and herbs.

*Evolution in pteridophyte and gymnosperm xylem: is it like
that in angiosperms?*—Occurrence of vessels has been considered a specialization in those few pteridophytes which
possess them. Likewise, Gnetales, the only vessel-bearing gymnosperms, are considered specialized in that respect. Beyond
this simple hypothesis, easily credible because of the systematic distribution of vessels within those groups, trends of
specialization have not been clearly outlined. White (1961,
1962, 1963*a*, 1963*b*) studied ferns with respect to tracheary-
element length and other features. Trends of specialization
did not emerge clearly from his work, for reasons indicated
later.

No phylogenetic series within gymnosperms based on features of wood have been proposed. The reasons for this go
beyond the relative uniformity of the ground plan of conifer
woods. Methods used in study of xylem phylogeny of dicotyledons and monocotyledons are, in fact, mostly inapplicable
to the xylem of pteridophytes and gymnosperms.

Data From Plant Physiology

Plant physiologists have appreciated certain xylem characteristics in designing their studies (e.g., gymnosperm wood
versus dicotyledon wood; diffuse porous dicotyledon woods
versus ring-porous dicotyledon woods). Plant physiologists appreciate that vessel elements represent an efficient system of
conduction (see Greenidge, 1957). Type of xylem is taken
into account in works of physiologists who have been concerned with rates of flow (e.g., Huber, 1935; Kramer, 1959;
Siau, 1971). However, the selection of species and genera
studied is not as large as one would wish, nor does it represent
many of the types on which one would like information because of anatomical characteristics. The comparative figures

that have been developed, however, are precisely in line with the theories—or better, syntheses—developed here. One may note that rates of flow have been measured in various ways: the speed with which a dye flows upward in a stem (Rivett, 1920); the speed with which a heat pulse applied to water in an intact stem flows upward (Huber, 1953); and the volume which can be forced through a stem of given transectional area during a given time utilizing known pressure (Peel, 1965). Obviously, comparisons can be made only within a single method of experimentation, although parallel results are, in fact, obtained.

Rates of transpiration are known for a wide variety of plants (Altman and Dittmer, 1966), both under field conditions and under experimental conditions. To be sure, many of the species studied are economic species. Even if these data are expressed in volumes of water transpired per unit leaf area, they are not necessarily applicable to the present study. Rates of transpiration are less directly related to xylem structure than rates of flow. For example, a species with a high transpiration rate may not have a high rate of flow in its xylem. Larger volumes of water may be absorbed by a larger root system—especially if the species grows in moist soil—and may flow slowly through a large transectional area of xylem. Even data on rates of conduction may not be conclusive. The fact that peak rates of conduction in conifers as high as those in ring-porous dicotyledons have been observed (Zimmermann and Brown, 1971) has been cited as a reason for disbelief in the efficiency of the angiosperm vessel (Bongers, 1973). The conifer conductive system is, to be sure, remarkably efficient, but particular sets of figures do not disprove the efficiency of the angiosperm vessel.

Scholander and his collaborators pioneered the measurement of negative water potentials in shoots and other plant portions by means of a pressure bomb (see, for example, Scholander, Hammel, et al., 1965). These negative water potentials do not relate at all directly to rates of flow. However, they do illustrate the magnitudes of tension of water columns within the xylem, and do illustrate that structural resistance to these tensions must play a part in xylem anatomy. Tensions can easily be so high as to shrink stem diameter measurably diurnally (MacDougall, 1921).

The only theory of water transport that accords with all known facts is the tension-cohesion theory of Dixon and Askenasy, best known in its statement by Dixon (1914). A historical review of this concept has been given by Zimmermann (1965); the essentials can be found lucidly explained in recent textbooks, such as those of Salisbury and Ross (1969) or Ray (1972). Many of the studies mentioned above have, in fact, had as their purpose the testing of this theory of water translocation, and comparative data are a mere by-product of that goal, since various types of plants ought to illustrate the tension-cohesion theory in various ways. Thus, some workers have deliberately worked with the most diverse kinds of plants (Scholander, Ruud and Leivestad, 1957; Scholander, Hemmingsen and Garey, 1961; Scholander et al., 1962; Scholander, Love and Kenwisher, 1955; Scholander, Hammel, et al., 1965).

The importance of data concerning the tension-cohesion theory of translocation is that if, as plant physiologists now assent, it is the only theory compatible with experimental data (Zimmermann, 1963; Scholander, Love and Kenwisher, 1955), it should be compatible with data from plant anatomy. This requisite is, in fact, satisfied to the extent that data from both fields are available, as will be shown below. However, a function of this book is in demonstrating the *ways* in which these correlations hold true. This book will also point out, if only inferentially, the species and plant portions for which physiological data are needed. As one can readily imagine, the most data have been gathered on species that are commercial timbers. Virtually no data are available, by contrast, for ferns.

One feature sometimes not appreciated is that only the periphery of a secondary xylem cylinder functions actively in conduction (Wray and Richardson, 1964), although inner portions of a trunk can be shown to function to a certain extent. The periphery is most active not merely because of the occlusion of heartwood tracheids or vessels by resinous materials, tyloses, etc., but rather because of the permanent development of air pockets ("cavitation," "air embolisms") in the water columns of plants (as stressed by Scholander, Hemmingsen and Garey, 1961). A large portion of older wood may be expected to be deactivated by this phenomenon. Kramer (1959, p. 655) states that "there is little doubt that an increasing fraction of the xylem elements of trees becomes filled with gas during the course of a summer . . . and in many species only a small portion of the total xylem is available for the ascent of sap." Not only does this constantly deactivate older portions of xylem, it must constantly be mended by active portions of xylem as well as, obviously, primary xylem of plants that have no secondary xylem. A particular problem is provided by long-lived plants that have no secondary xylem, such as palms (Ray, 1972), although we will see that there are explanations of how cavitations may be mended or prevented in all groups of vascular plants.

Also sometimes unappreciated is the fact that xylem offers resistance—often quite appreciable—to uptake of water, because of the capillarity or friction of tracheary-element narrowness, and because of the nature of their surfaces and pits. As Zimmermann (1965) notes, "If we put all this information together, we make the interesting discovery that trees require about 0.05 atm. m^{-1} to move water at peak velocities, irrespective of vessel diameter. This is because water moves faster in wide-porous species than in narrow-porous ones." Thus, slower translocation rates must characterize some species. Experimental data confirm this. For example, Huber (1956) gives the following rates for woody stems in cm^3 per sec.; conifers, 0.6; deciduous trees, 1.6–3.0; lianas, 7–36. Huber (1935) and Peel (1965) have found that conductive rates are more rapid in ring-porous woods than in diffuse-porous woods.

Wider vessels obviously have less friction and are suited to generally more rapid conduction.

Thus, we must answer why some woods with presumptively slow conductive rates are still extant. Is a slow conductive rate disadvantageous, and if so what compensations occur? Why haven't all plants developed the capability of conducting water as rapidly as lianas do? The hypotheses given below will attempt to answer such questions.

Data From Chemistry and Structural Engineering

Comparisons among species with regard to wood chemistry are not as useful as one might imagine, because such comparisons may not relate directly to structure and function. The most useful kind of data from plant chemistry is within a plant where comparisons to change in structure can be made. Such data, chiefly based on conifers (De Zeeuw, 1965) are available, and are cited below.

Comparisons of microstructures (e.g., pits) to macrostructures, as analyzed by structural engineering principles, may be questioned as being merely logical and not based on evidence. I feel such comparisons are more than mere analogies, however, especially in view of the fact that examples from microstructure and macrostructure so closely parallel each other.

New Data From Plant Anatomy

Development of some new data has been required for the goals of this book. I have not attempted to accumulate data on tracheids of ferns. These figures were developed in White's (1962) doctoral thesis, and I have used them here. Unfortunately, White published only condensed derivations from his original data.

Bailey and Tupper (1918) give data on tracheid lengths in gymnosperms, but their selection is relatively small. I have, therefore, developed new figures. In particular, gymnosperm woods I collected in Australia, New Caledonia, New Guinea, New Zealand and elsewhere have been particularly valuable because I knew from what portion of a plant each sample was taken, how old the plant was, as well as the size of the shrub or tree. This information proved essential.

The wood sample collection I have accumulated at the Rancho Santa Ana Botanic Garden has been invaluable for derivation of new data on gymnosperms and dicotyledons. Metcalfe and Chalk (1950) offer a wealth of data and their lists of families in which certain anatomical features occur are very useful. Mr. Larry De Buhr has given me his observations on secondary xylem of *Heliamphora* (Sarraceniaceae).

Since Cheadle's (1942) paper, new data on monocotyledon xylem have become available, some offered by Cheadle (1953, 1955, 1968, 1969, 1970; Cheadle and Kokasai, 1971, 1972). I have drawn on contributions by Ayensu (1972), Cutler (1969), Metcalfe (1971), Tomlinson (1961, 1969), and Tomlinson and Ayensu (1969). I have developed new data on Aponogetonaceae, Burmanniaceae, and Triuridaceae, and have been offered unpublished information on Triuridaceae by Dr. Margaret Stant.

Ecological Data

Inevitably, the hypotheses presented below require understanding of the water relations of innumerable species in their native habitats. Obviously such data are unavailable except for a few species. I have had the opportunity to observe many species in their native habitats, and I have drawn on these qualitative observations. The imprecise nature of these can easily be challenged. For example, I have observed that *Agave* and *Aloë* plants in culture at my residence rapidly absorb water during winter rains as evidenced by thickening of the leaves, but I have no quantitative data. I can, on the other hand, cite known data that *Agave* transpires at a rate slower than do nonsucculent angiosperms. *Agave* transpires chiefly at night, when its stomata are open, and leaf temperature and humidity minimize transpiration—a paradox permitted by its active dark carboxylation cycle (Neales, Patterson and Hartney, 1968).

With most angiosperms, transpiration is greatest when humidity is low and insolation is high (Gates, 1968). By increasing humidity, one can markedly diminish transpiration (Tinklin and Weatherley, 1968). Likewise, soil resistance is related to particle size, so that higher rates of transpiration occur in plants rooted in moorland soil, with intermediate rates in plants in sand and lowest rates in plants in clay (Tinklin and Weatherley, 1968). From experiments like these, one may, I feel, reasonably assume that transpiration is lower in plants of cool, humid upland cloud forest than in those of adjacent sunny, warm lowlands; and data to this effect have been developed (e.g., Coster, 1937). The extrapolation from data like those cited to the many species mentioned in this book may seem to involve unwarranted statements, excessively positive in tone. If so, I can only hope that data can be developed to further the knowledge of particular species.

For climatic data, I have consulted Walter and Lieth (1967). Their diagrams have the advantage of showing seasonal extremes. However, as any ecologist will agree, habitats within a given region can vary drastically in terms of water availability and transpiration. For example, wind is probably a very appreciable factor in xeromorphy of plants in the southwestern Cape region of South Africa, as are the highly porous sandstone-derived soils there.

In addition, different plants utilize the same habitat quite differently. As Kramer (1952) notes, "there are considerable differences among species in respect to root systems, dogwood being a notably shallow-rooted species, while pines often send roots to a depth of many feet." This may help to explain the paradox cited by Kramer (1952) that "conifers do not always have low transpiration rates and . . . sometimes lose as much or more water per tree as hardwoods of similar size." Caughey (1945) notes that coriaceous leaves do not always correlate with lowered transpiration. He finds that two leathery-leaved species, *Gordonia lasiantha* and *Ilex opaca*, transpire more per unit leaf area than poplars he studied. Plants with high transpiration rates may grow in highly mesic situations—but not always. The above problems should not be read as contradictions of the close interrelationship between ecology and xylem structure. Rather, we must understand the *total biology* of a species. *Allium*, a bulbous lilioid, may experience in its habitat near the soil surface a brief period of water availability and a long drought each year. Yet this *Allium* may be growing beneath a *Pseudotsuga* tree which taps a water supply throughout the year by means of deep roots.

The above factors will appeal to most readers as truisms that do not need repetition. I mention them only to show that discerning observation of species as they grow in their natural habitats is a prime desideratum. Extensive ecological data are unavailable at present. On the other hand, quantitative and qualitative knowledge of xylem anatomy can be obtained relatively easily in the laboratory. Therefore, I am forced to couple relatively precise anatomical details with vague ecological observations. Any further information will certainly be welcome, but I would hope that the totality not only of a plant's morphology but its ecology as well will be considered.

⨒2.

Ferns

The living ferns present interesting and subtle patterns with respect to xylem evolution, patterns that are perhaps best understood not in isolation from, but in relationship to fossil pteridophytes, living pteridophytes other than ferns, and even monocotyledons and gymnosperms. I am citing some of White's (1962) unpublished data and presenting new calculations based on those data. White's conclusions (1963*a*, 1963*b*) are entirely valid as far as they go. However, I find that the principles obtained from my survey of vascular plants as a whole require new constructions of those data. White focused on concepts of primitiveness and specialization, whereas I wish to trace selective influences.

THE SIGNIFICANCE OF TRACHEARY-ELEMENT LENGTH

Tracheary elements in ferns and the fibers associated with vascular strands are derived from the same procambial strands; or else the elongation in the two nearby tissue systems tends to be similar. Greater length of fibers (as in gymnosperms—see evidence cited in chapter 7) denotes greater mechanical strength of the plant portion. I find that length of tracheary elements in ferns is primarily influenced by what one could call degree of elongation of each organ in a fern. One could think of tracheary-element (and fiber) length as a function of how much the organ containing vascular tissue elongated prior to maturation and after transverse division of procambial cells had essentially ceased. In stems, one could think in terms of internode length, but that measure of elongation is inapplicable to roots and petioles. In White's work and the discus-

TABLE 6
Average Tracheid Lengths for Ferns

	Stems	Roots	Petioles
73 species with short erect rhizomes; terrestrial	1767 μ	4607 μ	5063 μ
29 species with long rhizomes; terrestrial	5503 μ	4271 μ	7334 μ
16 species epiphytes	2454 μ	1822 μ	3992 μ

(DATA computed from species means given by White, 1962)

sion below, the term "rhizome" is used to connote a stem, regardless of its form.

I have compiled (table 6) averages for roots, stems and petioles based on White's (1962) data. I extracted three groups, according to growth forms, and calculated data only for species in which White offered data for all three organs. In the 73 species with "erect rhizomes" (therefore less elongate than prostrate rhizomes) stem tracheids are clearly shorter, correlative to the acaulescent habit. Tracheids of roots and petioles in the 73 species are highly elongate in comparison. We can see this by comparison with the 29 species with long terrestrial rhizomes. In these, length of tracheids in stems far exceeds that in roots. A situation similar to the terrestrial species with long rhizomes is shown by the 16 epiphytes in table 6. One must remember that by far the majority of epiphytes have elongate rhizomes, and therefore should follow the pattern shown for terrestrial ferns with elongate rhizomes. The reason why epiphytes should have shorter tracheids than the terrestrial ferns with elongate rhizomes is that the epiphytes are generally smaller in plant size than terrestrial species within given genera or families. This tendency in turn is based upon the fact that plant size can be much greater in terrestrial situations where soil and shade provide greater and steadier water availability than does the epiphytic situation. Even within these groups, plant size correlates well with tracheid length. For example, in Grammitidaceae, a totally epiphytic family of small plant size, tracheids are short (*Adenophorus haalilioanus*: stems, 670 μ; roots, 790 μ; petioles, 990 μ; *Grammitis tenella*: stems, 855 μ; roots, 790 μ; petioles, 1,150 μ). In epiphytes of the Davalliaceae, which are larger plants, tracheids are longer (*Davallia dissecta*: stems, 3,655 μ; roots, 1,130 μ; petioles, 4,405 μ; *D. divaricata*: stems, 1,275 μ; roots, 1,440 μ; petioles, 5,620 μ). Probably the smallest of the epiphytes White studied is *Goniocormus minutus* (Hymenophyllaceae), and the stem tracheids (315 μ) are indeed the shortest for stems of any of the epiphytes.

Would an aquatic habitat provide longer tracheid length? Not if plant size is small. In fact, the two genera of truly aquatic ferns have exceptionally small plant size and appropriately short tracheids (*Azolla* sp.: 370 μ; *Salvinia* sp., 440 μ; these lengths based on macerations of entire plants).

White claimed that the reversal from shortest tracheids in stems to shortest tracheids in roots is correlated with phylogenetic advancement. This may be true, but if so, it reflects in fern families regarded as phylogenetically more specialized a tendency toward a higher proportion of species with elongate rhizomes. Epiphytes occur chiefly in Vittariaceae, Grammitidaceae, and Davalliaceae. Elongate rhizomes in terrestrial (sometimes epiphytic) plants occur chiefly in the families Gleicheniaceae, Marsileaceae, Polypodiaceae, and, to a lesser extent, Pteridaceae and Aspidiaceae. All of these families fall between 4.5 and 7 on the phylogenetic divergence (advancement) index (0 = most primitive, 10 = most specialized, based on morphological features) used by White (1962, 1963*b*) except for Gleicheniaceae (1.5). The remaining families all fall between 0 and 4 (exceptions: Parkeriaceae, 5.5; Blechnaceae, 5.5).

The tendency for stem tracheids to be somewhat shorter in erect stems was, in fact, demonstrated by White (1963*a*) in each of three species in which some individuals have erect stems, others procumbent rhizomes. Even in the latter, stem tracheids were still shorter than root tracheids in those three species. Noting the shift in stem tracheid length, I attempted to see if phylogenetic change in habit led to reversal in tracheid lengths in organs. To be sure, not all epiphytic ferns or terrestrials with elongate rhizomes have tracheids shortest in roots. For example, in *Dennstaedtia punctiloba* (terrestrial), tracheids average 2,835 μ in rhizomes, 3,960 μ in roots. However, the overriding tendencies do follow those illustrated in table 6. For example, the only species with root tracheids shorter than stem tracheids all have elongate rhizomes and are in Pteridaceae (*Histiopteris incisa, Hypolepis punctatum,* and *Paesia scaberula*). Most of the other Pteridaceae studied by White have very short, often nearly upright stems (e.g., *Adiantum, Pteris*). The same is true for the Aspidiaceae he studied, where all the species with tracheids shorter in roots than in stems have elongate rhizomes (*Cyclosorus goggilodus* [= *Thelypteris totta* and many other synonyms], *Dryopteris keraudrenianum,* and *D. novaboracensis*).

Tracheids in an acaulescent epiphyte, *Ophioglossum pendulum* (stem, 1,445 μ; root, 2,060 μ; petiole, 2,790 μ) deviate from the trends in table 6, but not greatly. The reason is the very short stem. In this connection, one may note that the very short stem tracheids of other Ophioglossaceae (*Botrychium matricariaefolium*, 735 μ; *B. pumicola,* 564 μ) conform not to concepts of primitiveness as related to tracheid length, but to the extremely short, small stems of Ophioglossaceae. Genera of tree ferns have relatively short stem tracheids (*Cibotium barometz,* 1,395 μ; *Sadleria pallida,* 1,630 μ; *Alsophila* sp., 1,765 μ; *Cyathea* sp., 1,120 μ). In this connection, one must remember that stems of tree ferns grow at a very slow rate; they are, in effect, rosette ferns that become epiphytes on their own dead stems, and stems are thus not really elongate in the same sense as those of rhizomatous ferns, which can have elongate internodes. Only a limited portion of a tree fern near its apex serves for water supply to leaves—only that portion which bears the living and active adventitious roots that grow down through the trunk into the soil. The peculiar pattern of tracheid lengths in *Oleandra* is dictated by the peculiar habit of this genus: elongate, "shrubby" stems, bearing leaves with short, stout petioles (plate 1-A). In *Oleandra costaricensis* average tracheid lengths are: stems, 4,365 μ; roots, 2,760 μ; petioles, 2,350 μ.

Longer tracheids (which tend also to be wider) could be said to have a conductive advantage of fewer, longer overlap areas per unit length of conductive tissue. This would make for more rapid conductivity per unit transection, as in conifer woods (Siau, 1971). However, shorter tracheids (which tend also to be narrower) would be less subject to collapse, for physical reasons, a feature that would be favorable where water tensions, an integral characteristic of conductive xylem of vascular plants, are high. If water tensions are not high in a given fern individual or organ thereof, long tracheids are not disadvantageous. One would expect a selective value for short tracheids in species of xeric localities or organs of species under high tension due to intense transpiration (or even occasionally so). Tracheid lengths in ferns (and other pteridophytes) would be expected to be a balance between these two opposed forces.

The factor of degree of organ elongation as a determinant of tracheid length seems overriding, but entirely compatible with the above possible factors. Long organs, which would tend to have long, wide tracheids, would not characterize xeric ferns; the ferns with large leaves are all highly mesophytic. As another example, the long, wide tracheids of fossil pteridophytes probably could not have existed under conditions of high water tension in the xylem.

However, short tracheids in *Azolla* do not connote xeromorphy. High conductive efficiency is not of positive selective value in *Azolla*, so short tracheids are not disadvantageous.

DIFFERENTIATION OF END AND LATERAL WALLS OF TRACHEIDS

If we look at White's (1963*b*) data on pits of tracheids of Pteridaceae, we find that in rhizome tracheids the pit width is about the same on end walls (1.23 μ) as on lateral walls (1.12 μ). In roots, the pit width increases on overlap areas between tracheids (1.75 μ) and coincides with decrease of pit width on lateral walls (0.75 μ). In the table for ferns given by White (1963*b*), there are 28 instances in which pits of tracheid overlap areas are wider than or identical to widths of pits on lateral walls, and only five instances in which the lateral wall pits are larger than those of an overlap area. White generalizes, in fact, by saying, "in each organ, rhizome, root, and petiole, the pits are smaller and areas of secondary thickening between the pits are wider on the lateral walls than on the overlap areas." This may represent a moderate differentiation in favor of greater conductivity, a sort of pre-vessel stage in tracheid specialization, in which pit areas are adjusted to the relative conductive performance of end walls as compared to lateral walls. The gain in mechanical strength of lateral walls by their lesser degree of pitting may represent a form of strengthening subsidiary to fibers. However, tensions in tracheids during conduction would dictate that resistance of metaxylem tracheids to collapse be retained by formation of no more pitting than is congruent with end wall and lateral wall conductive rates. Scalariform pitting in lateral walls of *Regnellidium* tracheids is abundant. This is comprehensible in terms of the permanently aquatic habitat of *Regnellidium*. Stems and roots are apparently perpetually immersed, and thereby high tensions probably would not develop in xylem and mechanical strength would not be a selective factor. However, there may also be a correlation with lack of sclerenchyma throughout the vegetative portions of *Regnellidium*.

VESSELS

Occurrence of vessels in ferns presents patterns which are at first confusing, but prove to be entirely understandable. Vessels are now definitely known to occur in roots, stems, and petioles of *Pteridium aquilinum*. This fern is definitely xeric in its preferences, so presence of vessels is indeed comprehensible, provided that one thinks of vessels as an accommodation to rapid flow of water when uptake is seasonally greater. *Pteridium aquilinum* does have aspects of xeromorphy. It survives drought without dieback of leaves. Gates (1968, p. 236) lists leaf internal diffusive resistance for 20 species. "Internal diffusive resistance" is a measure of combined resistance to transpiration from stomata, cuticle, and boundary layer. Of the 20 species of plants, *Pteridium aquilinum* has by far the greatest value for resistance, exceeding even conifers.

One might expect that seasonal fluctuation in water uptake in desert ferns would correlate with presence of vessels. Actually these ferns lack them, with the single exception, so far as is known, of presumptive vessels in roots only of *Notholaena sinuata* (White, 1963b). This is not really surprising, because roots and leaves of desert ferns die back rapidly during the dry season and the living portion of the plant survives merely as a small stem. The *Notholaena sinuata* pattern of vessels occurs (identification of vessels presumptive) in *Woodsia ilvensis* also (White, 1963b). In other species of *Woodsia*, tracheids with characteristics suggestive of tendencies toward vessel-like structure occur (table 10 in White, 1962). Presence of these vessel-like tracheids would correspond to the preference of *Woodsia* species for extreme localities with strong seasonal fluctuation (e.g., montane rocky outcrops), and thus the resemblance in vessel distribution and occurrence of vessel-like tracheids to the pattern of *Notholaena sinuata* is not surprising. In *Notholaena sinuata*, there is the possibility that vessels occur because roots convey water rapidly before onset of water stress and thus transmit water in a brief time to the stem. In

Marsilea, vessels are also known in roots (White, 1961). Here, roots absorb water rapidly prior to drying of meadows and ponds, so that growth and sporocarp formation are completed during the brief period of water abundance at pond and ditch margins. The lack of vessels in *Regnellidium* (Marsileaceae) would be explained by the fact that this Amazonian river-margin plant never experiences complete drying, only minor rises and falls in water level. Absence of vessels in *Pilularia* (Marsileaceae), which occupies sites like those where *Marsilea* is found, might be explained by the fact that plants tend to die entirely as soil drying occurs.

TRACHEID OVERLAP AREAS

The patterns of vessel occurrence of ferns would resemble those in monocotyledons, discussed in chapter 8, except that only a few ferns have vessels. A resemblance in organographic differentiation of tracheary elements like that of monocotyledons might be expected in ferns because both ferns and monocotyledons lack secondary growth and have adventitious roots instead of taproots. In fact, there is a diminished version of monocotyledons if we look at the portions of fern tracheids that overlap with other tracheids. White (1963*b*) presented data on width of bars and pits in the overlap areas of fern tracheids. White's data is presented in terms of average figures for pit and bar width for families; data were not presented in terms of species. Data were offered by White for widths of bars and pits on lateral walls of tracheids as well. The data on bars in the overlap areas of tracheids show no obvious significance in variation patterns. However, the data on pit width (and thereby, indirectly, contact area between tracheids adjacent in a vertical file) in overlap areas seem highly significant. If the figures for pit width are averaged from White's data, the results of table 7 are obtained.

The data of table 7 may not seem to show large differences, but because the pits are small, significance can lie in differ-

TABLE 7

Average Pit Width by Organs for Overlap Area Between
Tracheids in Approximately 120 Species of Ferns

Roots	Stems	Petioles
1.49 μ	2.03 μ	1.67 μ

ences of a few tenths of a micron (and such differences increase the pit area in, roughly, a square of the linear measure).
These differences are also significant because of the large number of measurements involved, and because the trends hold
in terms of individual families. In only two families, Cyatheaceae and Hymenophyllaceae, did White find that pit
widths were less in stems than in roots, and even in these two
families (which are highly mesic) the difference is extremely
small.

If wider pits in overlap areas are, for a relatively mesic group
of plants such as ferns, comparable to perforation plates, one
can claim greater conductivity and therefore accommodation
to moderate water-availability fluctuation primarily in roots,
secondarily in petioles. One might have expected this: roots are
adapted to conduct volumes of water, when available, more
rapidly than other organs of a plant without secondary growth,
judging from patterns in monocotyledons (see chapter 8).
Leaves of ferns are certainly subject to considerable transpiration. Unlike monocotyledons (which show considerable succulence or other forms of transpiration reduction in leaves,
adaptations to the sunny localities in which many monocotyledons grow) the broad surfaces presented by fern leaves are
antagonistic to minimization of water loss. Most ferns cannot
successfully minimize leaf surface and yet grow in the shady
localities they typically frequent. Therefore, moderate structural indication of accommodation to greater flow in petioles
might be expected. This might require, as a corollary, that the

conductively relatively "inefficient" stem tracheids (in terms of overlap areas) are more numerous to compensate, than petiole tracheids—or else that stem tracheids are of little importance because they are "bypassed" by adventitious roots. These questions need further study. In any case, the pattern of overlap areas in ferns in table 7 is simulated by distribution of vessels in the monocotyledon *Dracaena* discussed later, perhaps for similar reasons.

With respect to pit areas in overlap areas of petiole tracheids, three families run counter to the trend of table 7 by having the greatest pit width in petioles rather than in stems or roots: Pteridaceae, Grammitidaceae, and Marattiaceae. This may be correlated with rapid transpiration rates in leaves. Pteridaceae as a whole probably occupy sunnier sites than do other fern families. The Grammitidaceae studied by White are epiphytes subjected to bright illumination when cloud cover breaks. Marattiaceae do not follow these patterns, but the enormous leaf surfaces of *Angiopteris* and *Marattia* would be correlated with transpiration when humidity is lowered, whereas roots of Marattiaceae are always in wet soil.

A confirmation of the above may be found by consulting the portion of White's (1963*b*) table on fern families in which "end plates" present on tracheids (organs unspecified) are listed. An end plate may be regarded as a portion of a tracheid in which the overlap area is more discrete and offers broader contact area and therefore better conductivity in a file of tracheids. Of the Grammitidaceae and Vittariaceae White studied, 100 percent have end plates. This would be expected in these two families, because their epiphytic habitat may subject them to fluctuation in water availability and transpiration. Of Pteridaceae, 15 percent have end plates; 16 percent of Aspidiaceae and 16 percent of Marsileaceae have end plates. Because only families are designated, one cannot be certain, but one would expect that those species inhabitating seasonally dry localities are the ones that have end plates. More detailed data on end plates and pit areas of ferns would be of value, for indications from White's data seem clear when interpreted in terms of ecology.

TRACHEID WIDTH

Tracheid width might at first glance seem a potentially important factor in conductive efficiency in ferns. However, a check from my own slides of transections of rhizomes of five of the fern species studied by White showed diameter to be directly proportional to length. Wider tracheids would connote greater conductiveness. However, the amount of xylem in a stem must also be considered, so that tracheid diameter alone may not be a meaningful figure.

REVERSIBILITY

Does shortening of tracheids in ferns connote adaptation to xeric sites? White (1963a) showed that shorter tracheids— throughout the plants—occur in individuals of *Onoclea sensibilis, Dryopteris thelypteris* and *Woodsia ilvensis* from exceptionally dry localities as compared to tracheids in individuals of those species, respectively, from moist localities. However, White describes the plants from xeric localities as "dwarf," and this smaller size, as with the organ-length— tracheid-length correlation shown above, may be operative. Incidentally, the reverse trend was shown by plants of *Pteridium aquilinum* (longer tracheary elements in plants of drier localities). The reason for this cannot be determined without further examination of the response of this species, which could conceivably grow less well in shady wet localities. Tracheid length seems phylogenetically reversible in ferns; there is no reason to believe otherwise. Large and small plant and organ size seem reversible in ferns, and if size controls tracheid length, tracheid length is reversible.

However, vessel presence and degree of differentiation of

end walls of tracheids may not be reversible: studies leading to conclusions on this point are much needed. Also, physiological data are much needed in ferns: we know very little about uptake and transpiration in ferns.

❧ 3.

Pteridophytes
Other Than Ferns

EQUISETACEAE

Vessels have been reported in nine species of *Equisetum* (Bierhorst, 1958*a*, 1958*b*). In these species, vessel elements occur in internodal regions but not in nodal regions. Both aerial and rhizomatous stems contain vessel elements, but none have been reported in roots. Because Bierhorst found vessel elements in all the species he selected for study, we may assume that vessel elements occur more widely in the genus. We need to know if any species, in fact, lack them. Presence is not easy to demonstrate because vessel elements in *Equisetum* are tracheid-like, mostly with multiperforate end walls. This fact suggests that vessel elements are not markedly different from tracheids in their conductive capacity, and their occurrence should not be stressed highly.

Equisetum often grows in situations where roots are in moist soil, often at the edges of ponds, seeps, ditches, or in streambeds. These habitats provide relatively constant and abundant supplies of moisture to roots. However, the aerial stems of *Equisetum* are often brightly illuminated, and some are in warm regions where diurnal humidity drop may be considerable. This suggests that occurrence of vessels in stems might be advantageous during hours of rapid transpiration. In addition, shoot elongation in *Equisetum* may be rapid at the beginning of the growing season, and a slight increase in conductive efficiency by virtue of vessel presence might be a correlate.

SELAGINELLACEAE

Selaginella is known to possess vessels. Duerden (1934) searched for vessel elements in a wide range of *Selaginella* species, ranging from humid tropics to xeric mediterranean or desert climates. The former tend to be heterophyllous. The latter are mostly homophyllous, with scale-like leaves that inroll during drought. The homophyllous species of *Selaginella* in Duerden's study are characteristic of cliffs (often montane, with shady exposures, rocky ledges and rocky outcrops). These sites attenuate moisture availability compared to more exposed places in the dry regions where these *Selaginella* species occur. These rock crevices may be regarded as providing sufficient attenuation of moisture for existence of *Selaginella*, which otherwise would not occur in these areas. Even so, moisture availability for most of these species is highly seasonal compared with the tropical heterophyllous species, typically understory elements of tropical forests.

The homophyllous species studied by Duerden include *S. arizonica, S. bigelovii, S. densa, S. eremophila, S. hansenii, S. rupincola,* and *S. underwoodii.* Of these, all are restricted to the southwestern United States and adjacent Mexico except for *S. densa,* which ranges to the northwestern U.S., Manitoba, and the Alaskan panhandle. All of these possess vessels. All of them belong to *Selaginella* subgenus *Selaginella* section *Tetragonostachys* (Tryon, 1955). Another homophyllous species studied by Duerden, "*S. spinosa*" (*S. selaginoides*), does not have vessels. Not only does this species not belong to section *Tetragonostachys,* it occupies moister habitats: moist grasslands, hummocks, lake shores, and springy places across boreal North America, Europe, and Asia.

Study of section *Tetragonostachys* would probably reveal widespread occurrence of vessels, whereas mesic Selaginellas probably lack them. The occurrence of vessels in these xeric species is undoubtedly related to brevity of the growing season. Particularly in the southwestern U.S., rainfall may occur over a very short portion of the year; in some winters only one month may be moist enough to permit growth.

Although not conclusive, the illustrations by Duerden (1934) suggest that vessel elements of the xeric Selaginellas are thicker walled than are tracheids in the vesselless species. If so, this might reflect the occurrence of high tensions in xylem of the xeric species.

Of considerable significance is the fact that vessel elements in *Selaginella* almost all have simple perforation plates. This contrasts with ferns and angiosperms, where scalariform perforation plates give vessels a more tracheid-like configuration. The characteristically simple perforation plates of *Selaginella* suggest rapid conduction during the growing season, before drought causes leaves to inroll and the shoot, in effect, "closes down." One would guess that air embolisms may occur in vessels at the end of a growing season, but that repair of embolisms can occur rapidly (unless a particular shoot dies) when moisture is renewed. If so, this may occur by means of root pressure, because stems of the vessel-bearing Selaginellas are short and a modest amount of root pressure could fill vessels with sap again.

LYCOPODIACEAE, ISOËTACÈAE

Isoëtes, Lycopodium, and *Phylloglossum* lack vessels. *Lycopodium* species could all be said to be understory (sometimes epiphytic) elements in moist and humid forest, or else on slopes with considerable moisture availability owing to moisture seepage. These sites are certainly no less mesic than those occupied by vesselless ferns, which often grow with *Lycopodium. Lycopodium* seems clearly unable to enter habitats that could be called xeromorphic. To the extent species occur in more brightly illuminated places, leaf size and area are reduced proportionately. The inability of *Lycopodium* to exist in habitats as xeric as those occupied by *Selaginella* could be ascribed to lack of vessels in part, but other reasons seem more com-

pelling. For example, those species of *Lycopodium* with myco-
trophic perennial gametophytes are probably restricted by the
special boreal-forest habitats suited to slow growth of these
gametophytes. Those species of *Lycopodium* with photosynthetic gametophytes require moist conditions for gametophyte
development, and even these gametophytes may develop more
slowly than those of ferns—and certainly more slowly than
gametophytes of *Selaginella*.

The occupation by *Phylloglossum* of summer-dry areas illustrates by what remarkable modifications a lycopod is able
to exist in a mediterranean type of climate. *Phylloglossum*
grows in areas which are "temporary" bogs: wet in winter but
extremely dry in summer. The plant survives the summer in
the form of a "tuber" covered by suberized cells. Withering
of leaves occurs as soon as moisture becomes unavailable, so
that the actual growing period is no less mesic than in *Lycopodium*. The same could be said to be true, in an even more
extreme form, for *Isoëtes*. *Isoëtes* is typically a submersed
aquatic of ponds and lakes. Many ponds containing *Isoëtes*
dry completely during summer, but leaves have withered well
before the pond becomes dry. In these conditions *Isoëtes* over-
summers as a condensed stem that can be likened physiologically (but not morphologically) to the "tuber" of *Phylloglossum*. Freezing of ponds in winter leads to cessation of
growth in some *Isoëtes* species, such as the Tasmanian *I.
gunnii*. The high-Andean species which have branched stems
and form cushions, such as *I. andicola* and *I. gemmifera* (formerly segregated as a genus, *Stylites*) grow under conditions
of permanent water availability. The boggy lake margins
where they grow hover between drought by drying of soil and
physiological drought by freezing, but neither extreme is realized for protracted periods. In none of these circumstances,
then, is transpiration of a vulnerable leafy shoot active, and a
tracheidal conductive system is entirely adequate for transport
of water in the short stems of *Isoëtes*. The species with elongate stems, such as *I. andicola*, *I. gemmifera*, and *I. herzogii*,

have lateral adventitious roots that maintain a short distance of stem between point of root insertion and the leaf-bearing portion of the stem.

In mature *Isoëtes* plants, functional tracheids are almost all of the helically thickened type, despite lack of stem elongation and despite presence of secondary growth. This might relate to expansion and contraction of the stem itself during growth and dormancy, respectively, just as helically thickened vessels and vascular tracheids in Crassulaceae and cacti may relate to changes in stem volume during dry and wet seasons.

PSILOTACEAE

Psilotum and *Tmesipteris* are vesselless. This could perhaps have been predicted on the basis of xylem in the mycoparasitic monocotyledons, all of which are vesselless as far as is known. To be sure, Psilotaceae are holophytic, but as Bierhorst (1971, p. 155) states, "the major absorbing surface of the subterranean system is presented by the mass of fungal hyphae that occur within the cells of the cortex and extend out through the rhizoids and into the substratum." If mycorrhizae provide a slow but steady input of water, as hypothesized for the mycoparasitic monocotyledons, conductive rates rapid enough so that vessels would be advantageous cannot exist in Psilotaceae. Correlated with this would be the broom-like habit and minute scale-leaves of *Psilotum nudum*, which can exist in sunny sites. The epiphytic Psilotums (*P. complanatum, P. flaccidum*) have flattened axes (plate 2-C) which have a greater surface corresponding to broadened area for photosynthetic activity under shady conditions. These flattened axes are subject to greater transpiration than the terete stems of *P. nudum* (plate 2-B), but the epiphytic species grow in rather humid environments.

Also epiphytic in humid, shady situations (moist terrestrial in the case of *T. vieillardii*) are the species of *Tmesipteris*. These have "leaves" or pinnae relatively broad compared to the leaf-scales of *Psilotum*. These broader "leaves" are possible, with the conductive system of Psilotaceae, only because of their occupation of moist niches. Interestingly, the terrestrial *T. vieillardii*, which can grow on the floor of moderately dry forest as well as on the floor of cloud forest of New Caledonia has the greatest reduction in leaf surface for the genus. In *T. vieillardii*, moreover, "leaves" are oriented with edges above and below, so that minimal surface is exposed to sunlight.

The finite size of "fronds" in *Psilotum* and *Tmesipteris* may in part be dictated by limitations of the conductive system, which lacks secondary growth (plate 2-B,C), but mechanical considerations may be overriding, as suggested in chapter 5.

4.

Fossil
Pteridophytes

TRACHEID LENGTH

Despite technical difficulties, we know a good deal about vascular plan, degree of secondary growth, and pitting of tracheids in fossil pteridophytes. We know little about tracheid length. Bailey and Tupper (1918) studied slides and stated, "the tracheary elements in the secondary xylem in such forms as *Calamites, Sphenophyllum, Lepidodendron, Sigillaria, Lyginodendron, Heterangium* and other Cycadofilicales were undoubtedly very long, averaging several millimeters." We can assume that tracheids of these genera were probably longer than the average of tracheids of living ferns—perhaps they were comparable to the stem tracheids of ferns with elongate rhizomes (table 6). Living tree ferns, as we have seen above, have an entirely different structure, and are not comparable to fossil pteridophytes with secondary growth. If tracheid diameter in fossil pteridophytes is in proportion to length in accordance with the formulae of Bannan (1965), or something approaching those ratios based on gymnosperms, we could calculate tracheid length in fossil pteridophytes, and it would be very long.

If we except pre-gymnosperms (e.g., *Callixylon*, the wood of *Archaeopteris*), which functionally operate on the gymnosperm xylary pattern anyway, length of tracheids in the fossil pteridophytes does not connote a greater degree of strength. Tracheids of fossil pteridophytes are notable for large pit cavity and pit aperture sizes. If one views compilations such as those of Hirmer (1927) or Boureau (1964, 1967, 1970), one is struck by the weak tracheids and small amount of secondary xylem in these fossil plants, if secondary xylem is present at all. In fact, the problems of conductive tissue in those fossil pteridophytes that had secondary xylem are particularly interesting, and these have been selected for presentation.

Sphenophyllales

The stem of *Sphenophyllum* (fig. 2-A), like that of an *Equisetum*, had outer layers of sclerenchyma that provided support for the primary stem. However, this cylinder of sclerenchyma limits secondary growth, because no periderm replacement formed. The small amount of periderm formation in older stems appears to be of little or no mechanical significance. The primary xylem is triarch, and metaxylem tracheids had scalariform pitting. In all stems known, secondary growth is limited. In areas of secondary xylem alternating with the protoxylem poles, tracheid diameter was wide. This is suggestive of mesomorphy, and relatively good conductive ability. However, mechanical strength is limited.

However, some species, such as the *Sphenophyllum* of figure 2-B which had relatively thick-walled tracheids, showed indications of maximization of mechanical strength of tracheids. In that *Sphenophyllum* pits on tracheids were circular or polygonal in outline, with small apertures; these features increase strength, but the great abundance of pits does not (fig. 2-C). Circular or polygonal pits are stronger than scalariform pits, for reasons given elsewhere in this book. Ray cells in this *Sphenophyllum* (fig. 2-B,C) were thin walled. Ray cells were likewise of no mechanical value in *Sphenophyllum insigne* (fig. 3-A,B). In *Sphenophyllum insigne*, not only were tracheids thin walled (fig. 3-A), they also bore abundant scalariform pitting (fig. 3-C). Tracheids in secondary xylem of *Sphenophyllum insigne* increased the conductive volume of the

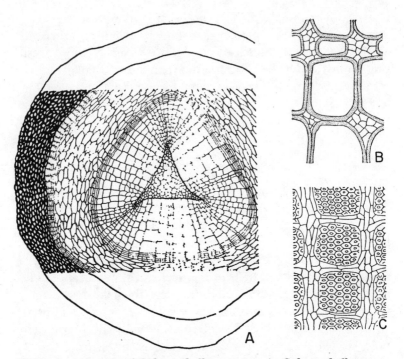

Figure 2. Anatomy of *Sphenophyllum* stems. A. *Sphenophyllum* sp., transection of stem showing secondary growth (drawn from Boureau, 1964). B. *Sphenophyllum* sp. (drawn from Renault, 1885), portion of transection of secondary xylem. C. *Sphenophyllum* sp. Portion of radial section of secondary xylem (after Renault).

stem, and seem maximally efficient for water transfer between tracheids, but they contributed little to mechanical strength.

Sphenophyllum was probably unable to increase mechanical strength concomitantly with increase in secondary xylem conductive capability. This may explain why *Sphenophyllum* was of limited stature. Stems barely attained 1 cm. in diameter, and plants were less than 1 m. tall (Hirmer, 1927).

Calamitales

Calamitales were the largest sphenopsids, with many species approximately 10 m. tall (Hirmer, 1927), and "possibly

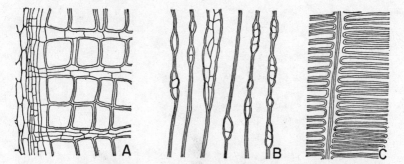

Figure 3. *Sphenophyllum insigne*, sections of secondary xylem (drawn from Williamson and Scott, 1894). A. Transection (cambium at left). B. Tangential section to show ray types (no pitting shown on tracheids). C. Portions of radial walls of tracheids, showing scalariform pitting.

achieving a height approaching 100 feet, although there seems no doubt the bulk of them were of much lower stature" (Barghoorn, 1964). The basic mode of stem construction, as shown by *Arthropitys communis* (fig. 4-A), features a sclerenchymatous periphery, like that of *Equisetum*. Barghoorn (1964) claims that in Calamitales, "support of the massive stems was achieved in large part through excessive periderm and secondary cortical expansion." As in *Equisetum*, vascular bundles of the primary stem were disposed in the form of a cylinder; this together with sclerenchyma and occurrence of nodal diaphragms, may explain the moderate, but obviously adequate, strength of primary stems.

The secondary xylem, however, was relatively weak in many species. The cylinder of secondary xylem was relatively thin (fig. 4-A), although the secondary xylem of Calamitales is claimed to have occupied a larger proportion of the stems than in Lepidodendrales (Barghoorn, 1964). In *Arthropitys*, tracheids were extremely thin walled, with scalariform pitting (fig. 4-B). Rays were non-lignified, cell walls thin (fig. 4-C). A secondary xylem with these characteristics would be expected to have limited capabilities for evolution into arbores-

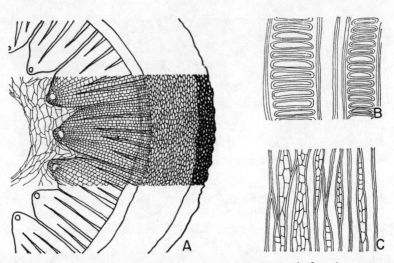

Figure 4. Anatomy of *Arthropitys* (Calamitales). A. *Arthropitys communis*, transection of a stem with secondary growth (drawn from Hirmer, 1927). B. *Arthropitys lineata* (drawn from Boureau, 1964), tracheids from radial section of secondary xylem, showing scalariform pitting. C. *Arthropitys lineata* (Boureau, 1964), tangential section of secondary xylem, showing rays.

cence. Secondary xylem accumulation was limited by the sclerenchymatous periphery. Breakage of this sclerenchyma by a periderm never occurred in Calamitales, apparently. Indeed, the intact sclerenchyma periphery of the primary stem may have been the mainstay of mechanical strength for calamitalean stems. This is suggested by the excellence with which external portions of calamitalean stems have been preserved.

The genus *Calamodendron* (fig. 5) illustrates an interesting essay in overcoming the mechanical limitations of the calamitalean stem. The cylinder of secondary xylem was not very thick. Rays were formed between each of the fascicular areas. Tracheids had scalariform pitting (fig. 5-C), but *Calamodendron* also had fiber-like tracheids with circular pits (fig. 5-B) in secondary xylem. The scalariformly pitted tracheids were thin walled, and were formed by the cambium in the central portion of each fascicular area (fig. 5-A). On either side of

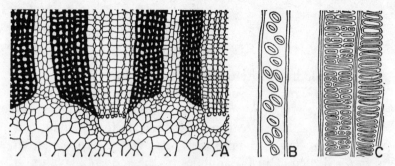

Figure 5. Anatomy of *Calamodendron* (Calamitales). A. *Calamodendron intermedium*, portion of transection of the cylinder of secondary xylem, with adjacent primary xylem (note lacunae) and pith at bottom (drawn from Renault, 1898). B. *Calamodendron punctatum* details (drawn from Boureau, 1964), portion of fiber-like tracheid from radial section, to show pitting. C. *Calamodendron punctatum* (from Renault), portion of conductive tracheid from radial section to show scalariform pitting.

these tracheids, the cambium produced bands of the fiber-like, thick-walled tracheids. Thus we see an alternative to the mechanism of production of earlywood and latewood tracheids by conifers. Growth rings of the type seen in temperate conifers, producing conductively efficient earlywood and mechanically effective latewood, were probably impossible in sphenopsids because of little variation in climate during the year. Calamitales probably existed in humid areas where terrestrial moisture was abundant, if not swampy, as indicated by the *Equisetum*-like morphology of underground stems. Under these conditions, and with linear or scale-like leaves of limited transpiration potential, slow conductive rates in limited amounts of secondary xylem probably prevailed in Calamitales.

LYCOPSIDA

Lepidodendrales

Lepidodendrales formed trees reaching 10 to 20 m. or more (Hirmer, 1927). This suggests that their mechanical systems were relatively effective. As shown for *Sigillaria* in figure 6-A, there are anatomical features that indicate the basis for this. The outer region of the primary cortex was sclerenchymatous, a fact reflected as in Calamitales by the good preservation of outer regions of stems of Lepidodendrales. The sclerenchyma of the primary stem is associated with the leaf cushions. This sclerenchyma provided aeration problems, and strands of parenchyma (parichnoi) extended from leaf mesophyll through the leaf cushion sclerenchyma into the stem cortex (Hirmer, 1927).

Primary xylem of Lepidodendrales was exarch and occurred in the form of a cylinder inside which was pith. This cylindrical construction might be thought to have maximized mechanical strength of the primary xylem, but was not of much

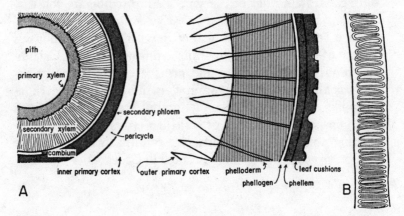

Figure 6. Anatomy of Lepidodendrales. A. *Lepidodendron vasculare*, diagrammatic transection of stem with secondary growth (modified from Hirmer, 1927). B. *Sigillaria bretoni*, portion of radial wall of tracheid, showing scalariform pitting (drawn from Boureau, 1967).

significance because of the smallness of the cylinder and the weakness of its component tracheids. Secondary xylem added relatively little to mechanical strength, because accumulation of secondary xylem was limited. More importantly, tracheids were thin walled and bore scalariform pitting (fig. 6-B) with narrow bars of wall material between the pits—mechanically a very weak wall. In some species of *Sigillaria*, scalariform pitting was strengthened by extension of fine network-like strands of wall material across pit apertures. Such networks ("Langsbälkchenstrukturen") were figured by Henes (1959, pp. 75–78) for *Sigillaria saullii*. They resemble the pits figured here for *Eupomatia laurina* (plate 11-E).

There was, however, a secondary cortex in Lepidodendrales. Often called a periderm, this periderm consisted of a very thick phelloderm—exceeding the thickness and total bulk of the secondary xylem. In fact, the drawing of figure 6-A represents rather more secondary xylem and less phelloderm than was typical of most lepidodendroid stems. However, both secondary xylem and periderm were thicker, pith thinner toward the base of a plant than farther up the trunk (Lemoigne, 1961). Phelloderm was sclerenchymatous, and thus undoubtedly did provide considerable mechanical strength. The periderm and primary cortex cylinders were interrupted by bands of thin-walled parenchyma in various ways, judging from drawings in paleobotanical literature. In any case, formation of secondary cortex did not break the primary cortex (although widening occurred at the trunk base). Thus, secondary xylem and secondary cortical (periderm) activity were confined within the rather rigid primary stem.

COMPARISONS OF FOSSIL PTERIDOPHYTES

The "closed" stems of Sphenophyllales, Calamitales, and Lepidodendrales, compared with secondary activity in conifers and dicotyledons, limited both the mechanical strength of stems and the amount of secondary xylem that could be interpolated into the confines laid down as primary stem. As Barghoorn (1964) said with regard to the conductive problems of the Lepidodendrales, "in relation to total size of the plant body, the giant lycopods seem to have achieved the extremes which have evolved in vascular plants with respect to the ratio of total mass of the plant to minimal development of secondary wood."

Such a system, moreover, could only be effective under highly mesic conditions. The histology and total amount of secondary xylem suggest volume of water conducted per plant per unit time was moderate. This, if true, is probably related to the limited laminar surfaces the Calamitales and Lepidodendrales had. Most species in these orders had linear leaves. *Sphenophyllum* had broader, thinner leaves, but it was an understory element. Various lines of inference suggest that Calamitales and Lepidodendrales were trees of marshy flats and grew in a humid tropical environment. Morphology of basal portions, small quantity of secondary xylem, and lack of growth rings all contribute to this conclusion. Even under these conditions, reduction in leaf surface was evidently a requisite for reducing transpiration to the conductive capacity of the secondary xylem.

In Sphenophyllales, Calamitales and Lepidodendrales secondary xylem mostly consisted of thin-walled scalariformly pitted tracheids: the "fern-like" tracheids often assumed to be ancestral to vessel elements in angiosperms (e.g., Bailey, 1944a). Although obviously the three orders of fossil pteridophytes have nothing to do with origin of angiosperms, they do demonstrate the nature of secondary xylem in which all tracheids are like those of ferns, and the limitations of that xylem. In no sphenophyll, calamite, or lepidodendroid genus do scalariformly pitted tracheids appear to have offered mechanical support sufficient for support except where the compensation of massive sclerenchymatous cortex and secondary cortex permitted the existence of a mechanically weak secondary xylem anatomy. Strong, thick mechanical cortex and the single thick stem characteristic of the fossil pteridophytes seem alien to any of the "primitive" dicotyledons, such as the vesselless species. These are all shrubs or trees, except for *Sarcandra* and the closely related vessel-bearing *Chloranthus*, and even in *Sarcandra* (plate 5-D,F) secondary xylem is mechanically relatively strong compared to that of the fossil pteridophytes. The vesselless dicotyledons and other dicotyledons rich in primitive characteristics are typical woody plants with bark quite unlike that of the fossil pteridophytes. As suggested later in connection with origin of vessels in angiosperms, secondary xylem like that of the fossil pteridophytes should not be hypothesized as ancestral to vessel-bearing dicotyledons. The files of fiber-like cells in *Calamodendron* show the sole solution—very unlike anything in gymnosperms or dicotyledons—to acquisition of mechanically strong cells in the secondary xylem of fossil pteridophytes.

One might invoke the hypothesis of paedomorphosis (Carlquist, 1962) to explain presence of scalariform pitting, like that of metaxylem, in secondary xylem of fossil pteridophytes. However, in stems of dicotyledons and cycads, there is an unbroken continuity between primary and secondary xylem. In the fossil pteridophyte orders considered above, however, Sphenophyllales and Lepidodendrales have exarch protoxylem. Thus there is a sharp break between primary and secondary xylem, although this is not true in Calamitales, which are endarch. However, exarch protoxylem also occurs in roots of dicotyledons, in some of which paedomorphosis could be said to occur. The point of significance seems to be that in the fossil pteridophytes pitting advanced to isodiametric bordered pits in only a few cases, whereas scalariform pitting predominated. The explanation of this would seem to be maximization of contact areas among tracheids and thus maximization of conductivity to compensate for limited transectional area of secondary xylem.

The invention of the gymnosperm plan of xylem, with its flexible combination of mechanical and conductive qualities,

would seem to be a prime reason for ascendancy of gymnosperms over the fossil pteridophyte groups. If so, this under-
lines the importance of xylem characteristics in evolution and
radiation of the major vascular plant groups.

Stelar Theory

Van Tieghem (1886) invented the term and concept of stele as a means of defining the limits of the vascular system and associated tissues. Having established these limits, Van Tieghem used the stelar concept as a means of categorizing types of vascular construction.

The chief botanist interested in exploring the significance of stelar types was Bower (1923, and his earlier contributions cited therein). Bower realized, for example, the role of cylindrical construction in the fibrovascular system of ferns (1923, p. 196): "no one can examine the form of the sclerotic sheath surrounding the dictyostele of Tree Ferns, and note the outward-curved lips round each foliar gap, without realising that it is constructed on the principle of a corrugated metal sheet, thus gaining great mechanical effect with little cost of material."

Bower notes the fact that young sporophytes of ferns have protostelic construction, which yields to medullated forms ontogenetically in some ferns. The greater complexity in vascular systems in ferns may take the form of polycycly (concentric rings of vascular tissue as seen in transection). This condition Bower (1923, p. 151) attributes to "the need for ready conduction along the axis, as well as to and from the enlarged pith." Dictyostely (p. 147) is, "a natural consequence of that shortness of the axis which is seen in ascending or vertical shoots, and of the crowding of the leaves which they bear." Foliar gaps, according to Bower, have the disadvantage of "laying open the conducting tract, and destroying the completeness of the endodermal control," but this is offset by the "advantage of ventilation" for medullary cells. Sclerenchyma, for Bower, is not merely of mechanical significance.

He states that "less definitely formed sclerotic masses may be found closely following the vascular tracts in stems and roots, but so disposed that . . . they act also as reservoirs for water of imbibition, as in the case of many xerophytic flowering plants."

The above conclusions seem tenable, although some may need restatement. Bower's strong concern with the importance of endodermis, particularly in the shoots of ferns, may be questioned. Regarding lack of endodermis in shoots of certain ferns, he says (1923, p. 185), "this state may serve for semixerophytic plants, such as the Ophioglossaceae and Marattiaceae, with sappy stocks and leathery leaves, and sluggish fluid transit. But it would not serve for plants where fluid transit needs to be rapid, and in particular where the leaf structure is delicate." There is no evidence yet concerning conductive rates in Marattiaceae, but the fact that in overlap areas of tracheids, pit widths are greater in petioles and stems than in roots (see chapter 2) does not suggest exceptionally slow conductive rates. Moreover, "leathery" leaves by no means always have lower transpiration rates than do thin leaves (e.g., Caughey, 1945). One may ask, moreover, how plants with the habit and the lack of secondary growth characteristic of Ophioglossaceae and Marattiaceae—namely, monocotyledons—can exist without stem endodermis. On the other hand, the leaves of *Pinus*, clearly xeromorphic, do have an endodermis around the vascular tissue. These examples illustrate that some correlations between structure and function are notably difficult.

At the risk of oversimplification, I would like to extend Bower's stelar concepts by presenting a series of principles regarding stelar structure primarily in pteridophytes, with comments on implications in other groups of vascular plants. Ferns have no secondary growth (except for very limited secondary growth in *Botrychium* and *Helminthostachys*), and have a conductive system of low mechanical strength. Nevertheless, principles derived from ferns are applicable to other groups of vascular plants. Although Bower's discussion is probably the most significant contribution on stelar theory, I have also drawn on data from Ogura (1972), who has offered a complete topographic survey of stelar types and purported phylogenetic relationships.

1. *The protostele is typically regarded by morphologists as characteristic of primitive plants, with abandonment of protostelic modes of construction as indicative of specialization. Yet protosteles occur in all groups of vascular plants, including angiosperms, in roots. The protostele has a functional explanation, and is a structural mode of optimal selective value in certain situations.*

The root, as cited, retains protostelic structure except where roots are wide and contain a pith (e.g., *Botrychium, Pandanus*). The universal presence of an endodermis in roots suggests control against outleakage of water from xylem and is also a means of canalization for conductive tissues. This control makes condensation of xylem and phloem into a central core, which can be surrounded by a minimum of endodermis, an optimal mode of construction. Tracheary elements grouped into a central core would be effective where root pressure occurs, and would not be disadvantageous under conditions of tension in water columns, where mutual support of adjacent tracheary elements would counter collapse. Aggregation of tracheary elements is not a matter of chance. For example, if distribution of water throughout a leaf were to be most efficient, veinlets would be one cell wide, but aggregations provide a better system where tension is involved. Aggregation of tracheary elements into strands also has significance in providing a redundancy of value if some tracheary elements in a strand are inoperative by virtue of development of air embolisms or other reasons; similar considerations might apply to sieve elements.

The central placement of a vascular core offers maximal protection of roots both from cortex-sloughing and damage of vascular tissues.

In an aerial axis, a protostele offers the potential advantage of photosynthetic tissue placement external to the core of tracheary elements. Where leaves are lacking, as they are in *Psilotum* (plate 2-B,C) or they were in *Rhynia* or *Horneophyton*, this is an optimal mode of construction. It ceases to be if the axis is relatively wide. An interesting example of protostelic construction within a photosynthetic cortex is provided by Orchidaceae which are leafless but have photosynthetic roots: *Polyrhiza*, etc. The pine needle provides another example, even, as in *Pinus monophylla*, with a single central strand of vascular tissue.

Even in leafy aerial axes, a central protostele provides advantages like those cited above if surrounded by an endodermis, as in *Lycopodium* or *Selaginella*. This is operative only where stem diameter is limited. For wider stems, some form of parenchymatization is characteristic, and thus a medullated protostele, discussed below, is present.

2. *Modifications of the protostele by means of lobing or irregular margins are chiefly related to appendages of an axis.*

As the generic name suggests, *Asteroxylon* was thought distinctive, compared with the other Rhynie chert genera, because of the stellate or lobed protostele as seen in transection. *Asteroxylon* is also the only genus of those three to have leafy appendages. The lobes seem clearly related to leaf traces that depart above the level of a given prominent lobe. Among contemporary lycopsids, this, as well as departure of adventitious roots, accounts for lobing (but not for formation of plates or medullation of the stele).

Roots of *Lycopodium*, *Selaginella*, and *Isoëtes* may dichotomize, but do not bear lateral roots. Steles of these roots may have two protoxylem poles—perhaps foreshadowing each dichotomy—but not a truly diarch condition with alternation of xylem and phloem poles as in roots of angiosperms.

The stelar plan of roots of seed plants is familiar: alternating xylem and phloem poles, surrounded by endodermis. If a large number of poles is present, there may be a pith. The number of poles is roughly proportional to root diameter. A question is posed by the occurrence of the alternating poles rather than collateral bundles as in the stem. Alternating xylem and phloem poles do form a pre-made system for connection of the vascular systems of a root with those of the lateral roots it bears. Lateral roots develop opposite the xylem poles, and rows of lateral roots corresponding to those poles can often be seen, especially on a plant in water culture. Collateral bundles in stems of seed plants may on occasion be connected to vascular tissue of adventitious roots, usually by de-differentiation of pith rays into procambium. The suitability of collateral bundles in the stem to their departure into leaves as leaf traces seems an overriding explanation for collateral structure in the stem.

However, the alternating nature of xylem and phloem poles represents another correlation. The exarch protoxylem poles are close to the endodermis, metaxylem elements farther away. In young root tips, protoxylem is active in the transfer of water from the root hair–cortex pathway upward. Metaxylem is matured at levels above active absorption by root hairs.

The plate-like protosteles of some *Selaginella* species can be related to heterophylly in that genus. The edges of the elliptical protostele (or meristeles) relate to parallel rows of leaves, and leaf traces depart from these edges.

In *Tmesipteris*, prominent lobing of the protostele is related to the traces that depart to "leaves." This is true in *Psilotum*, where the "leaves" are scale-like, but in that genus traces departing from the stele (plate 2-B) often relate to synangia, which are invariably innervated by a vascular strand.

Bower (1923) has pointed out that the endodermis in *Lycopodium* and other genera extends not only around the central stele, but around leaf traces as well. Presence of endodermis around leaf traces represents a canalization of conduction in which withdrawal of sap from conductive tissue by the leaf and subadjacent cortex may be hypothesized to be regulated so that constant but moderate water supply to the leaf is achieved.

3. *Modification of the protostele by means of a sclerenchyma sheath in the endodermal position serves a mechanical function, and control of lateral translocation from the vascular tissue as well.*

If sclerenchyma were formed in an axis solely for increase of mechanical strength—either self-support or increase in tensile strength—one would expect the most peripheral distribution of that tissue within the axis, unless the outermost region served for photosynthetic purposes. *Psilotum nudum* (plate 2-B) demonstrates this, although *Equisetum* is perhaps the most outstanding example. Collenchyma in various plant parts often does have the most peripheral position possible, and often occurs in the form of strands that interrupt peripheral photosynthetic tissue. Peripheral distribution of sclerenchyma does occur in stems of some species of *Lycopodium*, such as *L. pithyoides* or *L. phyllanthum* (plate 2-D). However, the majority of species of *Lycopodium*, such as *L. sitchense* (plate 2-E), *L. densum*, and *L. flabelliforme*, have a sheath of sclerenchyma around the stele. In these species, wall thickness of sclerenchyma cells is greatest adjacent to the stele, with thinner-walled cells adjacent to cortical parenchyma.

A sclerenchyma sheath substituting for a conventional endodermis is characteristic of a wide range of monocotyledon roots, as in Orchidaceae, Rapateaceae (Carlquist, 1966*b*) and palms (Tomlinson, 1961). The gross morphology of many of these plants suggests a mechanical function for roots. For example, sclerenchymatous endodermis is prominent in roots of epiphytic orchids but much less prominent or absent in roots of terrestrial orchids. The occurrence of sclerenchymatous sheaths as close to the vascular tissues as possible gives maximum protection from torsion, and compensates for mechanical weakness of tracheids and vessel elements. One cannot deny that the sclerenchymatous type of endodermis can serve the translocation control usually alleged for endodermis, but this seems secondary in such roots. As one might expect, sclerenchymatous endodermis occurs in roots that do not characteristically bear lateral roots—at least in the most sclerenchymatous portions. Such roots are not only elongate, they tend to be relatively long-lived (e.g., palms) compared to roots of bulbous monocotyledons. The mechanical nature of sclerenchyma in the endodermal position is confirmed when, as is often the case (e.g., "climbing roots" of lianoid aroids, etc.), mechanical tissue occurs within the stele also.

Worth noting, however, is Bower's (1923) attribution of endodermis presence throughout vegetative portions as a factor in the "triumph" of the leptosporangiate ferns. He claimed ability to deter outleaking of water from the vascular system to be the significant factor in the endodermis in fern stems and petioles. Outleaking would be of value if stems and petioles were photosynthetic, but they are not in many ferns, such as *Adiantum*, where ground tissue of stems and petioles is entirely sclerenchymatous.

Bower's hypotheses concerning endodermal function were advanced without benefit of modern concepts of translocation. We can now apply some of these. High tensions may develop in leaf tissues during periods of active transpiration, judging from the area of leaf surfaces of ferns. Presence of endodermis may prevent direct transmission of marked fluctuations in tension to xylem of stems and roots. If root and stem xylem has limited conducting capacity, presence of endodermis would tend to buffer conductive rates, assuring steadiness both in tension of the water columns and rates of translocation. Also, this would tend to prevent occurrence of high tensions within tracheids, which, if structurally fragile, might be subject to collapse. Measurement of tensions in fern xylem by means of methods such as those used by Scholander would be very informative.

In this connection, one may note presence of endodermis in conifer leaves (e.g., *Pinus*). Presence of endodermis here would be explainable as a mechanism for mediating transfer of water where differential tensions occur in leaf mesophyll and stem xylem. Since lowered humidity and prolonged bright illumination are known to raise tensions in leaves, either broad leaves in low humidity or acicular leaves in intense sunlight could have the effect of developing strongly fluctuating tensions, which an endodermis could buffer. This might also help explain presence of stem endodermis in certain broad-leaved dicotyledons, such as *Helianthus* and *Fitchia* (Asteraceae) and *Pentaphragma* (Pentaphragmataceae).

Where stem cortex and petioles are parenchymatous, complete sealing of the vascular system from outleaking would be disadvantageous. The fact that sclerenchyma is prominent in xeric ferns (Bower, 1923) may be explainable not so much as deterrence of outleaking from vascular tissue but rather as indicative of limitation of photosynthetic tissue to laminae. Laminae suffice for photosynthesis in these ferns. If chlorenchyma were present in stems and petioles of xeric ferns, they would be excessively vulnerable to desiccation. The only tissue with desiccation-resistance that can be substituted for thin-walled parenchyma is sclerenchyma, which is dead at maturity.

4. *The presence of parenchyma within protosteles (or other types of steles) in the form of occasional cells rather than as a pith may be related to water-conduction characteristics.*

Parenchyma cells scattered among metaxylem tracheids are characteristic of protostelic ferns (Gleicheniaceae, Hymenophyllaceae, Loxsomaceae, and many others), as well as Lycopodiaceae (species with actinomorphic protosteles) and subaerial axes of Psilotaceae. Examples of this type of parenchymatization are illustrated here in plate 1-B,C and plate 2-A,D. Such parenchyma is likely to be present where a broad area of metaxylem, as seen in transection, occurs. Scattered parenchyma cells in pteridophyte steles can be said to resemble "diffuse" parenchyma in secondary xylem of gymnosperms and dicotyledons, and may be present for similar reasons, such as lateral transfer of photosynthates and ions.

A protostele with occasional parenchyma cells among the metaxylem tracheary elements has been termed a "vitalized" protostele, but the literature gives no indication of why such "vitalization" should be of physiological value. Although the above explanation seems most likely at this time, one could also hypothesize that parenchyma cells adjacent to tracheids could prevent cavitation of tracheids by air bubbles, or even mend such cavitations. One may note that in all the major groups of fossil pteridophytes with secondary xylem, rays were present in secondary xylem. Such rays are illustrated here in figures 2-B,C; 3-A,B; 4-A,C; and 5-A. The fact that these fossil pteridophytes developed radial parenchyma systems would seem to favor explanations involving lateral transfer of ions or other solutes.

Interestingly, Osmunda (plate 1-D) is a fern with large masses of metaxylem tracheids lacking intervening parenchyma. In Osmunda, a dictyostelic arrangement provides, in effect, rays. The metaxylem strands are not as extensive as the metaxylem of the gleicheniaceous protostele, so introduction of parenchyma cells into the metaxylem of Osmunda seems unnecessary. The occurrence of plectostelic construction, as in Lycopodium (plate 2-E) may be regarded as an alternative to parenchymatization of a protostele. The plates of metaxylem in plectosteles are rarely more than four tracheids wide, usually less. The intervening bands of parenchyma between the vascular bands could serve for lateral translocation.

5. Broadening of the protostele occurs in accordance with broadening of the axis, as in Psilotum. This tends to place vascular tissue equidistant from the center of the pith and the outside of the cortex, which would have a physiological correlate. Mechanical factors are also involved in broadening of the protostele.

The degree of medullation, or pith formation, is almost per-

fectly correlated with axis diameter. This is most clearly seen
in roots of various vascular plants, but can also be seen in
Psilotum (plate 2-A,B) and *Tmesipteris*. The tendency for cylindrical disposition of xylem offers some degree of mechanical
strength, but rarely merely because of that placement alone.
Rather, placement of vascular tissue midway between center
and periphery of an axis is the most efficient way of assuring
minimal distance for translocation of water and photosynthates.

Mechanical strength is, in pteridophytes, almost always se-
cured by sclerenchyma. The cylinder of vascular bundles in
Equisetum would offer, by itself, only moderate increase in
mechanical strength; peripheral sclerenchyma and occurrence
of nodal plates achieve strength increase. However, placement
of bundles even in the *Equisetum* stem results in their being
midway between chlorenchyma and the edge of the hollow
pith. Increased height in *Equisetum* tends to be accompanied
by widening of the cylinder, as well as thickening of the cylinder, as in *E. giganteum*. This, in turn, is accompanied by
increase in sclerenchyma. The occurrence of T-shaped (in transection) sclerenchyma strands in interstomatal bands in *Equisetum* provides, in effect, a remarkable system of girder-like
members; the analogy here to T-shaped or I-shaped beams in
structural engineering is obvious.

Psilotum is instructive in that *P. nudum* has aerial axes of
appreciable size and weight, and (compared with the trailing
fronds of *P. complanatum* or *P. flaccidum*, plate 2-C) sclerenchyma is maximized. As shown in transection here (plate 2-B)
and in longisection by Bierhorst (1971, p. 164), the axes of
P. nudum have a central strand of "stelar" sclerenchyma as
well as cortical sclerenchyma underlying the chlorenchyma.
In macerations I have made, both cortical chlorenchyma and
non-photosynthetic parenchyma of all species of *Psilotum* and
Tmesipteris are fibriform, and their shape, toegther with relatively thick walls, provides additional mechanical support.

Where medullation occurs within a protostele, extensive

central parenchyma is not physiologically feasible (particularly where its outer limit is endodermal) without contact by means of a parenchyma zone with the cortical parenchyma. Leaf gaps provide one such contact. Alternative solutions include disappearance of pith (*Equisetum*) or conversion into sclerenchyma (*Psilotum, Matonia*).

Medullation of protosteles in roots is related to increase in root diameter. There can be roots with polyarch steles but little or no pith (*Marattia*), but pith is almost inevitable when there are approximately six or more xylem poles. Multiplication of number of poles in larger roots is associated with increase in absorptive capacity. Pith of polyarch roots is not isolated from contact with cortical parenchyma, although endodermis and pericycle intervene. However, pith of polyarch roots is often sclerenchymatous or relatively inactive, if not dead, thin-walled parenchyma. Polyarch storage roots have broad areas for contact between pith and cortical parenchyma.

Meristeles are often related to branchings. A pair of meristeles in a *Selaginella* stem can be considered a precocious dichotomy; however, at the point of branching of the stem, a cross-connection between the meristeles can often be found. Meristeles in *Lycopodium* are related to branchings. The flattened meristeles (which can thereby be considered plectosteles) found in *Lycopodium* stems often relate to dorsiventrality and occur in prostrate rhizomes. In these prostrate rhizomes, dichotomies occur not only laterally, but also there are dichotomies giving rise to an upright shoot and a continuation of the rhizome. These dichotomies are foreshadowed in configurations of the meristeles or plectosteles.

6. *The solenostele is a stelar type characteristic of rhizomes, and is related to leaves that are relatively large, but not as large as those of tree ferns.*

The fact that there is not much difference morphologically or functionally between a solenostele and a "vitalized" protostele is suggested by the fact that these types, respectively, can be found in two sections of the genus *Gleichenia* sensu lato (Bower, 1923).

Bower seems concerned by the potential lack of aeration of a pith enclosed in a cylindrical, endodermis-sheathed stele. There probably is little problem of aeration, even if a solenostele has both an external and an internal endodermis. The nature of the solenostele is merely a solution to the problem of supply of water and nutrients to both the pith and cortex, and its position midway between the center and periphery of a stem is an obvious correlate. The ferns with solenosteles have large parenchyma gaps where traces depart into leaves. There is, in fact, a subequal dichotomy of the solenostele because of the large size of leaves and leaf traces. The broad, long gap formed by departure of a leaf trace would afford considerable contact between pith and cortical parenchyma. The solenostele is correlated with long internodes; with short internodes, leaf gaps at any given point would be too numerous for reconstitution of an unbroken cylinder between the leaf gaps.

7. *Polycyclic steles are an extension of a solenostelic plan in which interpolation of additional steles is related to increased stem diameter or other factors.*

This principle can be shown in the successive sections Bower (1928) shows for *Pteris*. He also notes that *Dennstaedtia rubiginosa* has a central rod of vascular tissue within a solenostele, whereas *D. dissecta*, with a larger rhizome, has a second solenostele inside the first. In a genus closely related to *Dennstaedtia, Saccoloma elegans*, a broad upright rhizome is formed and this has vascular rods within what can be termed two solenosteles. *Matonia* has concentric solenosteles, and a moderately thick rhizome; there are ferns with rhizomes of approximately the same diameter but less complex vascularization. However, in *Matonia*, as in other examples, a greater degree of venation is achieved.

Polycycly verges on polystely, and resembles it in physiological effect of richer innervation of organs. Polycycly may be regarded as increasing complexity of vascularization in broaderrhizomed representatives of basically solenostelic fern groups, whereas polystely may be regarded as a similarly richer inner-vation in basically dictyostelic groups. A fact of importance in polycyclic ferns such as *Matonia* is that both external and internal solenosteles are involved in leaf trace formation.

8. *Dictyostely is, as Bower stated, a form of solenostely in which leaf gaps overlap; this, in turn, is related to condensa-tion of the stem, usually in relation to an upright rhizome.*

Beyond this simple statement, one can cite important fea-tures in the nature of leaf traces and leaf gaps of dictyostelic ferns. In the stele of *Stenochlaena tenuifolia* figured by Bower (1923), there are two structural factors worth noting. In this example, lateral traces as well as the main trace depart from a leaf gap and enter a petiole. In some dictyostelic ferns, such as *Bolbitis diversifolia*, figured by Ogura (1972, p. 56), leaf traces mostly depart from the lateral margins of the leaf gap rather than from its base.

A second feature of importance is that the main trace de-parting from a leaf gap may be dual; if single, it may be bi-furcated below. One extreme form that illustrates this (figured by Ogura, 1972, p. 71) is *Athyrium niponicum*: a pair of traces, one at each side of the gap, departs into a leaf. The duality of the main trace and the existence of lateral traces from the leaf gap indicate that the vascular supply of any given petiole is connected to opposite halves of the stele. Thus vascular supply assures delivery of water to a leaf and distribu-tion of photosynthates from a leaf in a way involving the broadest possible portion of the stele. This functional signifi-cance of the dual trace, basic to gymnosperms and angio-sperms, deserves emphasis.

Differences between the dictyostele described above and the dicotyledon eustele include: the occurrence of axillary buds and branch gaps in dicotyledons; the occurrence of clearly col-lateral bundles with endarch xylem in dicotyledons; the asso-ciation of extraxylary fibers at the protophloem poles of vas-cular bundles in dicotyledons; and finally, the absence of

endodermis in dicotyledon stem bundles. The basic similarities
are more compelling than the differences, and they indicate
parallel solutions of structural problems by ferns and dicotyledons.

Bower (1923) draws attention to the "perforations" (leaf
gaps) in the fern dictyostele, and implies a gain in "ventilation." He is preoccupied by separation of cortex and pith,
and views this as a "weak point" that is "set right" by "perforations" such as the leaf gaps in the dictyostelic ferns. The
questionable nature of Bower's preoccupation has been discussed above. Certainly in order to make statements like those
just cited, Bower would have had to believe that non-adaptive
types are in existence, a belief which seems to me to be untenable.

In eusteles of dicotyledons, trilacunar and multilacunar
nodes have the potential of connecting the leaf with a broader
portion of the stem vasculature than a unilacunar trace, and
the predominance of these types is understandable in dicotyledons. The monocotyledon node is, of course, a modified multilacunar type and thus has the potential advantage of drawing on the greatest possible proportion of stem vasculature.
There may be a relationship between this broad base of supply
for monocotyledon leaves and the longitudinally oriented nature of veins, with their relative paucity of interconnections
compared to the reticulate nature of dicotyledon leaves.

9. *In all ferns, sclerenchyma provides the chief source of
mechanical support. It may be entirely separate from the vascular system in locality and varied in quantity.*

To be sure, sclerenchyma often parallels the vascular system. In the case of pith sclerenchyma, one can hardly avoid
that interpretation, but here, as in other cases, sclerenchyma
is not derived from the same procambium that produces vascular bundles, as in angiosperms.

In no case, however, do the tracheary elements of pteridophytes supply major mechanical support, although White's
evidence, cited above, indicates a moderate potential gain in

mechanical strength where lateral walls of tracheids are sparsely pitted.

The most important implication, however, is that in ferns all mechanical tissue must be laid down in the primary body. This limitation is precisely that of monocotyledons. Interestingly, the range of growth forms in ferns is rather similar to that of monocotyledons. Lack of branching is discussed in chapter 8 for monocotyledons and those remarks are applicable to ferns. Monocotyledons have attained greater size in the cases of palms and vining forms than have ferns, probably because of greater conductive capability of vessels.

Mechanical strength can be increased or decreased by altering the proportion of sclerenchyma to thin-walled cortex. This is possible in cortex of dicotyledons, but the ground tissue of ferns and monocotyledons allows greater flexibility in increase or decrease in sclerenchyma.

10. *The monocotyledon stem and dicotyledonous stems with pith bundles represent a type of construction superficially like that of polystelic ferns, but in actuality these have been derived from different structural types.*

Monocotyledonous stems represent an inward displacement of bundles, compared to those of a typical dicotyledonous cylinder. As has been shown by various authors, bundles in monocotyledon stems move to a central position in the stem before turning sharply outward to enter a petiole (Zimmermann and Tomlinson, 1965). Thus, the most peripheral bundles represent leaf traces that will enter leaves far up the stem. Bundles are also more crowded at the periphery than at the center of the stem. These features, together with fibers associated with bundles, offer maximal mechanical strength potential. The numerous peripheral bundles also provide sites for the origin of, and connection with, a maximum number of adventitious roots, the sole kind produced by monocotyledons after the short-lived primary root has withered. Any given root supplies, thereby, not a trace to a leaf but a bundle likely, by branching, to supply several leaves in its upward course. The monocotyle-

donous stem may gain in flexibility by a sort of cable-like construction; this construction is pre-adapted to twining forms
and is advantageous in other types of construction as well,
as illustrated by palms.

The large number of bundles per stem and per leaf in monocotyledons provide a redundancy that compensates for failure
of any given vascular bundle. A vascular bundle with no secondary growth is more vulnerable to failure than one in which
new elements can be added. Multiplicity of bundles also compensates for the probability that lateral (tangential and radial)
transport in a stem is poorer in monocotyledons (although it
is possible via parenchyma) than in woody dicotyledons. De-
struction or deactivation of peripheral bundles is not fatal to
the plant.

Some dicotyledons have mimicked the monocotyledonous
plan—most notably *Gunnera*, which triggered Van Tieghem's
(1886) concept of polystely. There is little or no secondary
growth in bundles of polystelic dicotyledons such as *Gunnera*
or *Peperomia*.

Polystely in ferns is well represented by such a genus as
Marattia. Polystelic ferns may be hypothesized to have originated from dictyostelic types where broad stem and petiole
diameters require innervation by bundle-like segments of vascular tissue rather than arcs which would be isolated by relatively long distances from parenchyma if stem and petiole diameter were wide.

11. *Petioles mimic stems in solutions to mechanical problems, with differences based upon bilateral symmetry of the
lamina.*

As an organ with a tendency toward cylindrical structure,
this tendency in the petiole would be expected. However, the
petiole bears a flat, bilaterally symmetical lamina or series of
leaflets. Thus, the vascular and sclerenchyma systems of a fern
petiole are often semicircular or U-shaped, and the tips of this
arc depart as traces to pinnae. The U-shaped configuration provides maximal strength in a cantilevered position, in a situa-

tion where bundles depart and thus a complete cylinder cannot be achieved. Various modifications can result in increase in mechanical strength, and the variety seen in petiolar bundle configurations relate to compromises between strength and flexibility.

Some ferns, most conifers, Gnetales and ginkgophytes, as well as a scattering of dicotyledons, have a double leaf trace. Double traces have been figured for such ferns as *Onychium japonicum* and *Asplenium* sp. by Ogura (1972). A pair of bundles in the petiole, extending into the lamina before fusing, occurs in the cotyledon of many angiosperm species, and in the lamina of adult leaves of *Austrobaileya* (Bailey and Swamy, 1949), *Sarcandra* (Swamy and Bailey, 1950) and Chloranthaceae other than *Sarcandra* (Swamy, 1953). The basis of the dual trace, as stated earlier, is connection with both halves of the vascular cylinder. However, a potential disadvantage is that where two traces occur, interchange between the two halves must be achieved via parenchyma between the bundles. Transfer of water or solutes from one half of a leaf to the other is of value if one bundle fails or there is an inequality of conductive rates into the two bundles. This would explain why the dual condition is retained in adult leaves of only a small number of species, which occur under conditions where this is not disadvantageous. Duality is successfully retained in such conifer leaves as pine needles, where the terete nature of the leaf, the sheathing of the bundles by an endodermis, and the occurrence of transfusion tissue mean that the two bundles really do not function separately. Duality is also retained where a leaf appears to be a little-modified branch system, as in the case of *Stromatopteris* (Bierhorst, 1971, p. 199), where factors operative in branch systems can still be said to be current.

12. *Modification of vascular bundles and stelar systems relates to nutritional functions of the organs these systems supply.*

Predominance of phloem in vascular bundles connotes in-

put of photosynthates into an organ. This can be seen clearly
in floral parts such as nectaries (Frei, 1955) and various storage organs of vascular plants. If not actually more abundant,
sieve elements may show structure adapted to more rapid conduction of photosynthates (see chapter 8). Where water
translocation rather than photosynthate transport predomi-
nates, xylem outweighs phloem: hydathodes, for example.

Storage organs, by virtue of their width, tend to have some
form of alteration of the vascular system so as to innervate
massive zones of parenchyma. The monocotyledon stem is
ideally designed to be converted into a storage organ because
the scattered bundle system provides precisely the configuration fitting these specifications. Monocotyledon leaves also
tend to have more than a single plane of bundles when they
are thick and serve a storage function—either photosynthate
or water storage, or both.

The dicotyledon stem is pre-adapted to dispersion of bundles throughout a stem to the extent that it has numerous
bundles rather than, say, a solenostele. To be sure, drastic
modifications in bundle disposition can achieve dispersion
(e.g., potato, radish, kohlrabi). However, a notable innovation in dicotyledons that leads to interspersing of vascular tissue with storage parenchyma is the occurrence of successive
cambia or other cambial anomalies. Thus, *Beta, Boerhavia,*
etc., are well adapted to storage. This can be achieved by a
normal dicotyledon cambium if, in xylem, a large quantity of
parenchyma interspersed with relatively few tracheary elements occurs. This is characteristic of many dicotyledonous
stem succulents.

6.

Cycads and Ginkgo

CYCADS

Cycadales must be considered separately from other gymnosperms because their tracheids have no appreciable mechanical function. This is also true of *Welwitschia*, which, however, is not at all comparable in growth form or xylem to cycads. In xylem, cycads remind us somewhat of fossil pteridophytes, but their features are really quite different on account of growth form and habit. Although some cycads become small trees, all are relatively limited in stature and all can be said to be stem succulents. The abundant cortical and pith parenchyma of cycads is not water-storage parenchyma, but starch-storage parenchyma, so we may not wish to term them succulents; unfortunately, no truly appropriate term for this habit is available, unless we wish to invoke Corner's dubious and dubiously applicable term "pachycaulous."

Both the limited habit and the high degree of pith and cortical parenchymatization minimize requirements for mechanical strength and conductive efficiency. Fossil cycad groups agree with Cycadales in these respects, as well as in their probable tracheid length: 5,400 μ in *Cycadeoidea dartoni*, 5,100 μ in *C. dacotensis* according to Bailey and Tupper (1918). Although Bailey and Tupper presented data for mature wood of only one living cycad, *Dioön spinulosum* (see fig. 17), others may be presumed to be similar, for secondary xylem in the living cycads, despite certain distinctive features in some genera, conforms to the same basic plan (Greguss, 1968).

Secondary xylem of cycads accumulates extremely slowly and wide ray areas are maintained by the cambium. Rays are non-lignified. These conditions contrast markedly with wood

of conifers, taxads, *Gnetum*, *Ephedra*, and *Ginkgo*. Moreover, the pattern of tracheid-length change during addition of
tracheids (fig. 17) does not show a precipitous shortening
from protoxylem to metaxylem (compare *Pseudotsuga*, fig.
17-A). Increase in length of tracheids is very gradual after the
onset of secondary growth. Tracheids are relatively wide in
diameter and do not show any growth-ring phenomena. Because of stem succulence, as well as reduced leaf surface, one
can hypothesize that water stress does not govern patterning
of secondary xylem. One can say that cycads have "internal
mesomorphy," that is, because of succulence and reduction in
transpiration, water is conducted in small volumes and probably very slowly. Measurement of water tensions via the
methods of Scholander, as well as conductive rates within the
stem, would be very desirable, although technical problems
when dealing with cycads would be great.

Because of lessened mechanical stress compared to woody
plants, tracheid abundance and strength are not of strong
selective value. The great length of cycad tracheids, comparable in that respect to those of a tall conifer, may seem at first
paradoxical. One could say that because xylem is accumulated
so slowly, the relatively few tracheids that are formed should
be longer, and therefore both mechanically stronger and conductively more efficient. However, a form of juvenilism, or
paedomorphosis, is operating. In vesselless woods, there is a
marked upswing of tracheid length at the beginning of secondary growth compared to dicotyledon woods with vessels (compare *Pseudotsuga*, fig. 17-A, to the curves of fig. 15). This is
due to greater elongation of the fusiform cambial initials themselves (not their derivatives) in the vesselless woods. This
tendency occurs precociously in cycads because of the very
slow accumulation of xylem. However, this trend does not
continue—the "record" of tracheid lengths in a radial section
shows that tracheid length does not continue to increase and,
in fact, decreases slightly. This must be due to occurrence of

transverse (pseudotransverse) divisions, the kind that occur during increase in girth of the cambium. These divisions also occur in dicotyledonous woods that show paedomorphosis, such as rosette trees and stem succulents, in such a way that element length decreases or stays virtually the same. In cycads, one can hypothesize that increase in diameter of the xylem cylinder occurs slowly at the onset of secondary growth. Therefore the transverse divisions that in a woody stem would occur frequently, decreasing length markedly during the transition into secondary xylem, do not occur frequently in a cycad, so that average length of initials is cut down slowly, while the tendency for apical elongation of the cambial initials themselves, mentioned above, occurs.

Paedomorphosis is also indicated in cycad woods by the fact that scalariform lateral wall pitting, characteristic of metaxylem elements, extends indefinitely into the secondary xylem in many species (Greguss, 1968). This is correlated with the fact that secondary xylem does not serve for mechanical strength in cycads. Self-support of cycad stems is achieved by the large parenchyma zones in cortex and pith, as in dicotyledonous stem succulents such as cacti. Some cycad species do develop, in outer wood, the tracheids with circular bordered pits characteristic of conifers.

Cycads have fewer tracheids per unit area of stem transection, or even per unit area of xylem transection, than do conifers. This is characteristic of a succulent mode of xylem construction (see table 13). There is some irony in the fact that cycad leaves are not like those of a leaf succulent; they are definitely xeromorphic, however, compared to those of ferns. Cycad leaves have very thick cuticles and sunken stomata, overarched by subsidiary cells. The least xeromorphic leaf anatomy can be found in *Bowenia*, which is an understory element in tall rain forest of Queensland.

GINKGO

The close similarity in histology of reproductive stages between cycads and *Ginkgo* dramatizes their quite different
vegetative morphology and anatomy, as well as different
sporophyll nature. *Ginkgo* follows the same principles in wood
construction as conifers (Greguss, 1955), and subsequent remarks about anatomy of conifer and taxad wood are, in general, applicable. *Ginkgo* is unique compared to conifers in that
it is both broad-leaved and deciduous. We do not know for
sure the native habitat of *Ginkgo*. However, it is clearly temperate, whereas the broad-leaved conifers tend to be from
humid tropical and subtropical areas. In these respects, there
are parallels between *Ginkgo* and *Tetracentron* among the
vesselless dicotyledons. We can hypothesize, as for *Tetracentron*, that *Ginkgo* requires mesic summer conditions. This
remark applies not to cultivated species of *Ginkgo*, for *Ginkgo*
can be successfully grown in a wide variety of temperate
localities. However, *Ginkgo* reaches great age and reproduces
spontaneously only under conditions with moist spring and
humid summer conditions, such as Japan. The limited distribution of *Ginkgo* prior to presumptive extinction of wild
populations may be due to its highly specialized ecological
requirements, and also to the fact that suitable habitats support a large number of contenders among woody dicotyledons.
These hardwoods were presumably less abundant during the
times when Ginkgophytes were more abundant, as shown by
the fossil record.

Conifers and Taxads

All gymnosperms are vesselless except the three genera of Gnetales. Because of presence of vessels in *Gnetum*, *Ephedra*, and *Welwitschia*, they are discussed in chapter 10 in connection with origin of vessels in dicotyledons, although there is no phyletic connection between Gnetales and primitive dicotyledons.

SIGNIFICANCE OF TRACHEID LENGTH

The pattern of vesselless conifers (Araucariaceae, Cupressaceae, Pinaceae, Podocarpaceae, Taxodiaceae) and Taxaceae (*Cephalotaxus* should perhaps be regarded as a separate family) proves to be a simple one. We must view tracheid length and tangential diameter separately from radial diameter of tracheids, however. Tracheid length represents, with a small degree of fluctuation (mostly due to intrusive growth of tracheids as they mature) the length of fusiform cambial initials. Tangential diameter of tracheids is also controlled by tangential width of fusiform cambial initials. Radial diameter of tracheids can, however, change during maturation, often quite markedly. Wall thickness of tracheids can also be altered without regard to dimensions. One should note that there is a relationship between tracheid length and diameter (Bannan, 1965). This ratio conforms to a curve, with diameter predictable on the basis of length (length-to-width ratio increasing with increased tracheid length). This ratio is applicable only to tangential diameter of tracheids, a fact of which Bannan is aware.

With respect to tracheid length in conifer stems, the factor of prime importance is clear: length is directly related to height of plant or size of branch (if the wood is from a branch) and to subsidiary growth-form characteristics. In order to establish that this is, in fact, the basic factor, I calculated tracheid length for 120 species of conifers and taxads, many of which were species not included in the Bailey and Tupper (1918) survey. One expression of my results can be found in the top bar of figure 11. Bailey and Tupper (1918) stated, "the smaller, slower growing forms, e.g., certain Taxaceae and Cupressaceae, tend to have shorter elements than larger, more rapidly growing forms." As a result of more extensive data, I can report that the relationship is a very clear one. It is sometimes obscured if, for example, one has wood from a small stem of a species that eventually develops a large trunk. In that case, one finds tracheids shorter than one would expect, judging from the size of a mature tree. If one notes the nature of length-on-age curves, as shown for *Pseudotsuga* (fig. 17-A), one sees a marked increase in tracheid length, then a leveling-off. A wood sample from a tree 25 years old would obviously have shorter tracheids than a sample from a tree 100 years old. The length of tracheids in the 25-year-old tree relates very directly to the stature of that individual. The flattening of the curve demonstrates optimal size, so that the height of the trunk and the weight of branches have become constant. As new branches beyond this point are formed, an approximately equivalent number die.

The Mechanical Hypothesis of Tracheid Length

The tracheid-length–tree-size correlation implies that longer tracheids provide greater strength (although correlative features of wide, long tracheids are integral, as will be seen). Let us suppose that mechanical strength is the prime factor for tracheid lengths. In this case, one might ask why, once maximum size has been reached, "release" does not occur, for the bulk of accumulated wood should suffice (if wood stays in a healthy condition, as it may not) for mechanical strength, and tracheid length might even shorten. Of

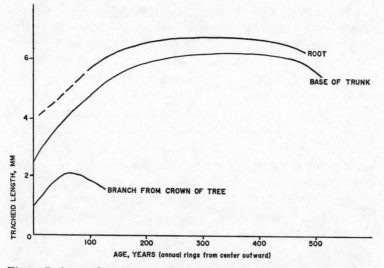

Figure 7. Age-on-length curves for tracheids from different portions of trees of *Sequoia sempervirens*. Broken line at beginning of root curve is a projection; data begin at solid line. (Modified from Bailey and Faull, 1934.)

course, even if the quantity of branches stayed the same, each year would add more weight. Banks (1973) has shown on the basis of structural engineering principles that the factors influencing trunk strength in a tree are resistance to wind-thrust and support of the individual itself. In fact, tracheid length may well eventually shorten in very old trees, as suggested by the curves of figure 7 for *Sequoia sempervirens*. One may guess that in many conifers, attainment of maximum size is followed soon by death of the tree, so that there is little or no opportunity for release from the positive value of mechanical strength. The nature of tracheid features in the outermost wood of an old tree gives us, in fact, several criteria, or "controls," to show that there is still a requirement for mechanically strong elements. If the requirement for mechanical strength lessened, one might expect not only shorter tracheids; one would certainly expect production of limited numbers of tracheids (or of radially very wide tracheids that would be more efficient conductively than narrow tracheids), and the production of weaker, or of thin-walled, tracheids as well. Actually none of these changes occurs to any marked extent, indicating that a steady requirement for mechanical strength exists. The same can be claimed for vessel-bearing woody dicotyledons, where if requirements for mechanical strength lessened at maturity, the proportion of vessels to imperforate elements could easily be changed, as could wall thickness of imperforate elements. However, the few changes that have been detected from the innermost to the outermost wood in dicotyledons indicate that no alteration in favor of lessened mechanical strength occurs (see data from Sastrapadja and Lamoureux, 1969, cited in chapter 1).

In developing data on tracheid length in conifers and taxads, I knew the approximate age of the samples I had myself collected, and the size of the plant or branch from which a sample had come. In dealing with cut wood-blocks in wood-sample collections, one lacks this information. Consequently, if one took data from wood-block samples, and compared tracheid length to typical height for mature trees of those respective species (as given, for example, by Dallimore and Jackson, 1966), one would find discrepancies. For this reason, the tracheid lengths given by Bailey and Tupper (1918) are not reliable indicators of mature tree size for a species; where their figures fall short, their samples probably came from immature trees.

Bannan (1965), in his review of variability of conifer tracheid length, cites such ecological factors as increased aridity or increased altitude of habitat or the stunting effects of winds as influencing decreased length of tracheids. If one takes into account the fact that the individuals in these suboptimal situations are smaller, the correlation of length of stem tracheids with size of plants remains excellent. I might mention that with respect to average tracheid length of species, I found both greater lengths and shorter lengths than those reported for the assortment of species studied by Bailey and Tupper (1918). Tracheids of *Agathis palmerstoni* (7,740 μ) and A. *vitiensis* (7,650 μ) are notable in this regard. The species in which I found notably short lengths can be reliably reported, because with small plant size, one can easily collect a sample from the base of a plant. Species worthy of mention here include, in Cupressaceae, *Diselma archeri* (699 μ); in Podocarpaceae, *Microstrobos niphophilus* (721 μ), *Dacrydium laxifolium* (941 μ), *Microcachrys tetragona* (1,126 μ), and *Phyllocladus alpinus* (1,134 μ); and in Taxodiaceae, *Athrotaxis cupressoides* (1,750 μ). Interestingly, these all come from extremely cold Southern Hemisphere localities: montane Tasmania and New Zealand. Also notably short in average length of tracheids are species of the genus *Actinostrobus*, a cupressaceous genus endemic to Western Australia: A. *arenarius* (1,674 μ); A. *pyramidalis* (1,727 μ), and A. *acuminatus* (1,757 μ).

An interesting feature with respect to tracheid length is the tendency for greater length in conifer roots than in stems. This was reported by Bailey and Faull (1934) for *Sequoia* (fig. 7), and has been confirmed for other conifers by Bannan (1965). Bannan is undoubtedly right in implying that in the two cases where Fegel (1941) obtained contradictory results, unrepresentative samples may have been studied. One might conclude that greater length of tracheids in roots implies greater mechanical strength. However, one must keep in mind that in roots, tracheids tend to show less fluctuation in radial diameter and have less marked growth rings than do stems for any given species; root tracheids correspond more to what one might term earlywood tracheids. Root tracheids would tend, by virtue of length and wideness, to have greater conductivity. Greater length would provide longer overlap areas on which more pits could occur, and greater length also provides fewer tracheids per unit length of wood, and thus less impedance to conduction according to the considerations of Siau (1971). Wider radial diameter provides greater area for wider pits on overlap areas. Whatever their mechanical characteristics, root tracheids would therefore appear to have greater average conductivity compared with stem tracheids. Wide, thin-walled tracheids could be tolerated in roots if such tracheids (which are vulnerable to collapse where high negative tensions in water columns are present) do not experience high tensions in roots. The Scholander considerations cited later suggest that within a tree, negative pressures are greatest at the top of a plant, least at the roots.

Upper branches (as opposed to those near the ground) tend to have long tracheids (Bailey and Tupper, 1918; Bailey and Faull, 1934). In this connection one may note that branches are relatively exposed, and that the mechanical strength of the main trunk is enhanced by a greater number of years of xylem accumulation. Branches from the crown of an old tree, however, may have relatively short tracheids (fig. 7). This would correlate with the fact that a small crown bears a limited amount of foliage, and is subject to less mechanical stress than a large branch.

A drop in tracheid length with injury has been noted in conifers (Bailey and Tupper, 1918). This seems inevitable, for regrowth of cambium over an injured area represents an interruption of the long initials on an uninjured trunk, and short tracheids tend to be produced in wood with irregular grain, as at the base of branches.

In support of the mechanical hypothesis, one can cite experimental and chemical work. Wardrop (1951) and Wellwood (1962), both working with conifers, found that there is an increase in tensile strength with an increase in tracheid length and corresponding fibril angle, from the pith outward. There is a similar increase, paralleling increase in tracheid length, in proportion of cellulose in cell walls from pith outward in conifer stems (Wardrop, 1951; Hale and Clermont, 1963). There is a similar increase in cellulose crystallinity from pith to periphery (Lee, 1961). Increasing cellulose crystallinity corresponds to increasing shear resistance and tensile strength (Mark, 1965).

The fact that greater tracheid length, as noted for conifer roots above, connotes greater conductive efficiency, creates difficulties in separating which factor is of prime importance in determining tracheid length. However, if greater tracheid length were determined by conductive efficiency alone, one might expect greater lengths in conifers that transpire more actively; or, more significantly, conduction of greater volumes of water per transectional area per unit of time. We have no measurements done comparatively and under controlled conditions, unfortunately. If one could judge from known ecological preferences of species, one might say that, considering conifers of both markedly xeric and markedly mesic sites, the tracheids are neither longer nor shorter than what one would predict on the basis of plant size alone. Worth noting, nevertheless, is the fact that earlywood tracheids tend to be shorter than latewood tracheids (Spurr and Hyvärinen, 1954). This suggests that elongation to achieve greater mechanical strength is a factor of prime importance within growth rings; if earlywood tracheids elongated as much, they probably would not increase their conductive capacity, because addition of pit areas probably would not occur in proportion to degree of intrusion. Even in roots, one would have to explain why growth rings occur at all if water-conduction factors are of prime importance. Roots of conifers may, however, experience higher negative pressures during the time of year when latewood is formed, and this may influence tracheid diameter.

TRACHEID DIAMETER AND ITS INTERPRETATION

Bailey (1958) has stated that, "the two major functions of tracheary cells are to a considerable extent antagonistic, since certain structural features that enhance rapid conduction tend to weaken the cells, whereas others that strengthen them tend to retard the movement of water from cell to cell." As we have seen, if we consider length alone, mechanical strength of tracheids does not run counter to conductive efficiency, but parallel to it. However, one could well ask why, if long tracheids are advantageous, all conifers do not have long tracheids. A short tracheid does not appear to be advantageous by virtue of length alone, but rather because it tends to have a narrower diameter also. As we will see, there are situations under which narrow diameter, and thus shorter length also, is, in fact, advantageous for conifers.

The antagonism between conductive efficiency and mechanical strength is certainly obvious when we examine the effect of tracheid diameter. For greatest mechanical strength, a tracheid should have a thick wall, a narrow lumen, fewer pits, and pits smaller in diameter. All of these features run counter to conductive efficiency.

In viewing the effect of diameter, one must remember the effect of transectional area. For example, I measured earlywood tracheids of *Taxodium mucronatum* (plate 3-C) as averaging 37 μ in diameter (lumen width). This provides a transectional area of 1,369 sq. μ, compared with lumina averaging about 180 sq. μ or less in latewood. In contrast, the stem of *Microstrobos niphophilus* (plate 3-D) showed growth rings consisting mostly of latewood tracheids, the lumen area of which averages about 8 sq. μ. *Taxodium mucronatum* is a tree of stream beds in subtropical Mexico. *Microstrobos niphophilus* is a shrub from windy montane areas, frozen in winter, of Tasmania. *Taxodium* earlywood ought to be suited to more rapid flow, whereas *Microstrobos* must have a very slow rate of flow. Compared with vessel-bearing angiosperms, conifers have typically been found to have slow rates of conduction (Epstein, 1972; Huber, 1953, 1956).

Wall thickness change within a growth ring connotes division of labor. Earlywood tracheids are weaker because of larger and more numerous pits, as well as because of thin walls. Latewood tracheids are stronger because of narrower diameter and thicker walls, but also fewer and smaller pits. In addition, there is more cellulose—a criterion for mechanical strength—in latewood tracheids of conifers (Ritter and Fleck, 1926). Division of labor between earlywood and latewood tracheids minimizes the "antagonism" that Bailey (1958) saw as an inherent limitation of the tracheid.

The example of *Taxodium mucronatum* above reminds us that the tangential diameter of tracheids is fixed by the tangential diameter of fusiform cambial initials. The radial diameter can vary, but within certain limits. Root tracheids can be radially very wide. This is especially notable in submersed roots of *Taxodium distichum*, which grows in lakes and swamps. It is also true in two riparian conifers from New Caledonia, *Dacrydium guillauminii* and *Podocarpus minor* (plate 3-B). In these two species, the widened trunk bases, as well as roots, have tracheids which may reach 50 μ or more in diameter. Some of these tracheids are wider radially than tangentially. From my observations, I would guess that radial diameter rarely exceeds 1.5 times the tangential diameter in conifer tracheids.

In tropical conifers, radial width of stem tracheids tends to fluctuate less than in temperate species. In *Agathis* (plate 3-A), *Araucaria*, and broad-leaved species of *Podocarpus*, wood may appear weak on account of the consistently wide tracheids. This is probably more apparent than real, for the average wall thickness proves, when measured, not to be lower than in temperate conifers.

Radial widening of tracheids has only a limited effect on lessening friction in water passage through tracheids, when one views them in terms of the lessening of friction obtained by the degree of wideness of angiosperm vessels. However, one must remember that vessels do not comprise the entire axial mass of a dicotyledon wood, whereas tracheids do comprise the conductive cells of a conifer wood.

Narrowness of tracheids does have an advantage: resistance to the negative tensions in the water columns. Wide elements

would collapse more easily than narrow elements under strong
tension, for reasons to be found in structural engineering (for
example, the narrower a tube or pipe, the thinner-walled it
need be to withstand positive or negative pressure). Scholander, Hammel et al. (1965) have shown that negative pressures of up to —60 atm. can occur in juniper (presumably
Juniperus occidentalis), the greatest negative pressure they
measured except for mangroves and *Larrea divaricata*. They
found negative pressures of up to —15 atm. at the top of
Sequoia sempervirens trees, and up to —25 atm. at the top
of *Pseudotsuga menziesii* trees. Strong negative pressures must
be resisted by mechanically strong cells. That this requisite is
a reality is demonstrated by collapse of wood tissue in natural
stands of white spruce (Lutz, 1952). This occurrence was
apparently the result of warm weather that induced transpiration at a time when roots were still frozen and water was
relatively unavailable. The very strong negative pressures in
juniper are probably tolerated by virtue of the wood structure
characteristic of *Juniperus occidentalis*, that is, narrow, thick
walled tracheids. Such tracheids are also short, in accordance
with Bannan's (1965) correlations. Within the conifer trees
studied by Scholander, Hammel, et al. (1965), the highest
negative pressures were found in twigs from tops of trees.
Tracheids in the twigs would, in accordance with the tracheid-length data cited at the beginning of this chapter, be
relatively narrow and short. If negative pressures are consistently low, one would expect that "release" would occur,
and that longer, wider thin-walled tracheids would be formed.
As suggested earlier, this appears to be exactly what does occur
in conifers. Such "release" could also occur in conifer species
that never experience extremely high negative pressures, and
this may explain why the longest (and therefore also the
widest) tracheids thus far measured occur in *Agathis*, despite
the fact that *Agathis* trees, although quite tall, do not equal
Sequoia sempervirens in maximal height.

These considerations suggest that three major factors operate in both parallel and contradictory fashion with respect to gymnosperm tracheid characteristics: resistance to negative pressures; mechanical strength for self-support of the tree and resistance to wind thrust; and area of tracheid overlap and its pitting, which determine conductivity. In a tree that experiences moderate negative pressure in water columns, tracheids can be long, wide, and have relatively thin walls; thus the pattern of *Agathis australis* (plate 3-A), or *Podocarpus minor* (plate 3-B), or the pattern in roots of many conifers can be achieved. An increase in seasonality places maximum value on conductivity during the season of active growth and transpiration; in a basically mesic situation, this results in long tracheids that are wide and very thin walled in earlywood, a condition that requires compensation in latewood by tracheids that are radially narrow and thick walled (and incidentally more resistant to negative pressures). The pattern of *Taxodium mucronatum* (plate 3-C) and many other tall temperate conifers illustrates this condition. Where negative pressures are very high and transpiration is low, and thus conductivity value is lowered, tracheids that are short, narrow, and thick walled are to be expected. This pattern is seen in *Microstrobos niphophilus* (plate 3-D); the shorter tracheids are not disadvantageous if mechanical strength is not of strong value, and thus shrubby conifers follow this pattern. The *Microstrobos niphophilus* pattern is simulated by young trees and by the uppermost branches of tall trees, where very high negative tensions develop and the value of mechanical strength is not great, but in these cases seasonal demand for higher conductive rates is accommodated by wider, thin-walled tracheids in earlywood, narrower thin-walled tracheids in latewood. Obviously these "equations" do not include many subtleties of habitat or of the compensatory or modifying effect of foliar type.

EVOLUTIONARY SIGNIFICANCE OF CONIFER WOOD

If we seek the closest parallels between wood of conifers and taxads and that of angiosperms, we find similarities at opposite ends of angiosperm specialization.

The closest parallels to the woods of conifers and taxads can be found, among angiosperms, in the vesselless families (plates 4-7). The vesselless angiosperms have limitations, on account of their xylem plan, similar to those of conifers, a situation discussed in chapter 9. Next in degree of similarity to conifer woods would be vessel-bearing angiosperms in which vessels are primitive and tracheid-like, such as *Illicium* (plate 9). Dicotyledons with vesselless woods or woods like those of *Illicium* probably have slow conductive rates and are limited to mesic situations: moist soil and humid air, conditions under which high negative pressures probably do not develop in xylem. Among conifers, *Agathis* and some species of *Podocarpus* (e.g., *P. blumei*) follow this pattern and are discussed below. Some conifers have been reported to have conductive rates so slow that they cannot be measured (Parker, 1956); measurement of conductive rates and sap tension in broad-leaved tropical conifers is very much needed.

Shrubby xeromorphic and alpine conifers can be likened not only in external appearance but in certain wood characteristics to a high-alpine Andean shrub, *Loricaria thuyoides* (Carlquist, 1961b). The axial portion of the xylem in *Loricaria* consists of large numbers of extremely narrow vessels and vascular tracheids (plate 15-C). These tracheary elements are very short (plate 15-D). No fibers are present. Thus, if we neglected the presence of perforation plates on vessels (vascular tracheids are vessels so narrow that they do not bear perforation plates and are thus designated by that rather confusing term), the wood resembles that of *Microstrobos*. The very narrow vessels and vascular tracheids suggest occurrence of high negative pressures in xylem, as well as slow conductive rates—conditions that occur in juniper (Scholander, Hammel, et al., 1965). The wood features of *Loricaria*, if plotted on the graph of figure 14-B, would fall at the upper end of the curve drawn. Incidentally, the species of that graph at the upper end of the curve are desert shrubs. If gymnosperm tracheids were considered vessels for the purpose of that graph, they would represent the ultimate upward extension of the curve—extremely narrow, with numerous elements per unit area.

Microphyllous conifers can and do exist in alpine regions like those inhabited by *Loricaria*, the numerous alpine conifers of New Zealand and Tasmania being examples. Conifers can be said to be microphyllous in the northern deciduous forest where winter temperatures are below freezing for months of the year. At such times, conifers transpire almost as little as bare branches of deciduous trees (Kozlowski, 1943). The deciduous *Larix* is leafless during below-freezing weather. In summer, ground water is available in the northern coniferous forests, so that even if transpiration rates are moderately high (Kramer, 1952), summer survival is assured. One would expect that in below-freezing conditions in the northern coniferous forests, when water is much less available and, according to reports, water columns in wood are largely frozen, transpiration rates of conifers fall to virtually zero.

Conifer leaves have exceptionally low transpiration rates because they have very high "internal diffusive resistance" in leaves (Gates, 1968, p. 236). This measure, a composite figure for the tendency of leaves to resist desiccation per unit of surface, shows conifer leaves transpire less than any of the angiosperms listed by Gates. High diffusive resistance is related to leaf anatomy. For example, stomata of conifers are typically plugged with "alveolar occluding material." (For a review of literature on this phenomenon, see Wulff, 1898.) Jeffree, Johnson, and Jarvis (1971) have shown that because of stomatal plugging, transpiration rates are lowered by two-thirds, while photosynthesis is lowered by only one-third in *Picea sitchensis*. This is a remarkable parallel to occlusion of stomata in Winteraceae and Trochodendraceae. In addition, overtopping of stomata by subsidiary cells minimizes exposure of stomata to the environment. Conifer leaves have thick cuticles as well. The presence of endodermis around bundles, as in *Pinus*, further restricts water withdrawal from leaf xylem. Even with these features that assure high diffusive resistance, microphylly in conifers is extreme. One can say that further reduction in leaf surface might lower photosynthetic capability below the compensation point. Because of extreme reduction in leaf surface and therefore volume, compensation in photosynthetic surface by means of small intercellular spaces and infolded walls of chlorenchyma cells (*Pinus*) seem inevitable results.

Many conifers survive in xeric situations. Low soil moisture can be compensated for by a deep root system or the ability to "close down" roots during dry periods by means of corky or sclerenchymatous coverings of roots.

A microphyllous woody species with a slow rate of conduction and a long life cycle and therefore slow rate of replacement can still be quite competent in certain sites. Conifers are not often weeds, but *Pinus pinaster* is a noxious weed in the Cape region of South Africa. Africa south of the Sahara is devoid of native Pinaceae. *Pinus pinaster*, the Mediterranean pine, can succeed in sandstone soils of mountains in South Africa because it is adapted to the high acidity of those soils, it is efficient at drawing on water deep underground in these porous soils, and it can stand soils poor in certain minerals. Ironically, the veld fires of South Africa burn shrubs of open country to the ground. However, *Pinus pinaster*, in contrast to the *fynbos* vegetation, is tall and crowns of the pine trees are untouched by the fires, which stay close to the ground and sweep through areas rapidly. Pines are also ideally suited to resistance to the high winds frequent in the Cape region. Perhaps because in no other place (except where Pinaceae are native) do such unusual ecological features exist, the conifers here are predisposed to weediness.

Reversibility

An interesting feature concerning gymnosperm wood is that while irreversible trends of wood evolution have been proposed for dicotyledons, no such ideas have been formulated for conifers and taxads. In fact, there is no reason to believe that there has been progressive phylogenetic decrease in length of fusiform cambial initials (and thereby tracheid lengths) in gymnosperms as has been hypothesized for dicotyledon woods.

Conifers are commonly thought to have become restricted to, or to be optimal only in, high latitudes or high elevations or both, and that this restriction has become progressive as angiosperms have radiated and displaced them. However, two of the largest genera of conifers have probably been able to invade the tropics to a limited extent. *Podocarpus* (ca. 100 spp.) is notable in this respect. Some *Podocarpus* species, such as *P. blumei*, which is broad-leaved and grows in tropical lowlands with typical tropical trees such as Dipterocarpaceae, are tall trees and have appropriately long tracheids. *Podocarpus blumei*, *P. vitiensis*, and *P. wallichianus* have broad leaves with flabellate venation rather than the single vein typical of most species of *Podocarpus*. This suggests phylogenetical broadening of the lamina. These broad-leaved *Podocarpus* species (subgenus *Nageia*) have cone structure regarded by Florin (1951) as not primitive but highly specialized for the genus. *Podocarpus* has probably radiated in austral forests because of its ability to mimic in habit and leaf size and shape the broad-leaved dicotyledons of these forests. These dicotyledons have slow rates of replacement, so the long life cycle and slow rate of replacement of *Podocarpus* are not disadvantageous in these sites.

Juniperus (ca. 60 spp.) has radiated and probably escaped the restrictions of boreal conifer genera such as *Larix*. Some *Juniperus* species are large trees, but *Juniperus* has not invaded the tropics to a spectacular extent. However, *Pinus* (ca. 55

spp.) has managed a notable extension into tropical areas, chiefly tropical uplands. The tropical pines are striking in tallness and in great tracheid length, and include such species as *P. hondurensis*, *P. merkusii*, and *P. roxburghii*. In the sample of *P. roxburghii* I studied, tracheids averaged 7,578 μ in length.

The occurrence of a large (150 feet tall) species of *Callitris* (Cupressaceae), *C. macleayana* (average tracheid length, 4,905 μ) in the wet forests of Queensland suggests entry into more mesic sites by a family that one associates with xeric habitats or strenuous temperate regimes.

Interestingly, *Podocarpus*, *Juniperus* and *Pinus* have all been able to establish on subtropical oceanic islands. This suggests their pioneering ability, capability for adaptation to more moderate climates, and relative ease (for conifers) of dispersal. All of these insular circumstances are like a replication of the invasion of tropical regimes by these groups.

On the other hand, one can hardly doubt that Araucariaceae is a subtropical family of a relictual nature, as the distributions given by Florin (1963) suggest. Taxodiaceae are also notably relictual, and seem to have been unable to remain in, or re-enter, subtropical or tropical regions. That Taxodiaceae were once much more widespread is suggested by fossil distributions and by marked disjunction of the single genus in the Southern Hemisphere, *Athrotaxis*, restricted to Tasmania.

Many botanists seem to be of the opinion that angiosperms have supplanted gymnosperms because of the efficiency of the angiosperm life cycle. I feel that the advantages of the angiosperm life cycle are very real. Vesselless angiosperms probably did gain a foothold by virtue of relatively rapid reproduction, and perhaps also an ability to occupy understory situations in conifer forests. The nature of the conductive system of angiosperms has probably been of greatest importance in the rise and diversification of angiosperms, not merely because it is "more efficient," but because in its varied evolutionary patterns, angiosperm xylem lends itself to the widest possible range of habitats occupable by vascular plants. The severest limitation of gymnosperms is not so much in geographical. range, but in diversity of habits (for life cycle reasons, gymnosperms cannot be annuals, for example) and habitats. There is an interesting confirmation of this described below, the similar restriction of the vesselless dicotyledons. Also the restriction of woody dicotyledons with long scalariform perforation plates to mesic sites and the "relictual" nature of those species is yet another example of the limitations a xylem formula can impose.

Function Histology of Conifers and Taxads

Pits of coniferous tracheids are most abundant on radial walls, often very scarce on tangential walls. The radial walls are, when seen in tangential sections, to a large extent overlap areas between any given tracheid and those adjacent to it. Thus, radial walls are maximized as conductive areas. Radial walls that face rays are abundantly pitted (see Greguss, 1955). The large size and abundance of tracheid to ray pits suggests what Braun (1970) terms "contact cells," although evidently gymnosperms, in Braun's ray taxonomy, are regarded as having "isolation cells." The great length of conifer tracheids makes contact with at least one ray by any given tracheid virtually inevitable. Axial parenchyma cells are relatively abundant in some conifer woods, such as *Agathis australis* (plate 3-A), where they appear mostly at margins of rays, or *Podocarpus minor* (plate 3-B).

There are several interesting consequences of the above familiar facts. Conduction by any given tracheid is not isolated from conduction by adjacent tracheids. Thus, the entire outer sheath of secondary xylem in a conifer stem or root conducts water as a continuous cylinder. This contrasts with vessel-bearing angiosperms, where uptake occurs almost exclusively within a network of vessels within the outer portion of the secondary xylem. In these terms, conifers seem almost more efficient than angiosperms. Although vessel-bearing angiosperms are typically more efficient than conifers in rate of flow (Huber, 1953, 1956), rarely less so (Zimmermann and Brown, 1971), conifers can equal angiosperms in transpiration per plant (Kramer, 1952). The operation of the total outer cylinder as a conductive zone in conifers permits such a paradox, and helps explain how conifers can be competitive.

Bordered pits in conifers were shown by Bailey (1958) to have visibly porous pit membranes when carefully studied with the light microscope. These micropores in conifer pit membranes have since been elegantly and clearly demonstrated by electron microscopy (e.g., Liese, 1965; Tsoumis, 1965; Siau, 1971). These porosities occur in the margin (margo) of the membrane. The circular thickening familiar as the torus occupies the center of the pit membrane. The porosities in the pit membrane explain the relative conductive efficiency of the conifer tracheid, even if far less in conductive rate potential than a vessel element. The torus is of significance in closure of a pit by means of its valve-like flattening against a pit border, as shown by various authors (e.g., Bailey, 1958; Tsoumis, 1965), thereby deactivating a tracheid.

The pit border cannot be considered a structure of value only as a portion of this valve mechanism. The bordered pit occurs in angiosperm tracheary elements, where tori are absent. The most important function of the border is minimizing interruption of the secondary wall—and thereby retaining wall strength—while maximizing, by means of the wide pit membrane, conductive contact between tracheids. Water transfer through membrane porosities would have to equal water flow through the pit apertures for this to be operative.

Scholander, Hemmingsen and Garey (1961) characterize cavitation (occurrence of air pockets within water columns) as frequent, and state that flow of sap by means of root pressure is insufficient to mend this "gas seeding." They note that mass cavitation of wood occurs when water in xylem freezes during the winter. In the case of conifers, however, this may not permanently deactivate xylem. Zimmermann (1964) states that vessels of angiosperms are permanently deactivated each year unless—as occurs in certain species—a rise of sap in the spring displaces the gas in vessels (in a manner not yet described). However, Zimmermann (1964) notes that the porosities in pit membranes of conifers, because of their small diameter, are "permeable to water but do not let an air-water interface pass." Therefore, in conifers, "when the tree freezes, each tiny bubble remains confined within a single cell, and when the water begins to thaw, the gas bubbles are sufficiently dispersed so that many of them can redissolve before tensions are developed by transpiration." If so, this opens the possibility that a rather wide cylinder of sapwood might be functional in conifers. In angiosperms, isolation of cavitation is much more difficult, because when vessel elements occur in a long vertical file, cavitation anywhere within the vessel could quickly spread to the length of the vessel. Thus, conifers possess a potential advantage over angiosperms in this respect. At the very least, we can see why vesselless conducting systems have proved phylogenetically viable. Angiosperms, as we will see, have mechanisms for repair of, or bypassing of, cavitations. In gymnosperms, there is a possibility that parenchyma might aid in repair of cavitations. All conifer woods have rays, and few lack axial parenchyma (Greguss, 1955; Jane, 1956). However, axial parenchyma is at best a small proportion of the xylem; it tends to occur in latewood, often, as figured by Esau (1965, p. 251). Very likely the main function of parenchyma in conifer woods is as a secondary conductive system, as appears to be true in dicotyledons (see chapter 11). The relative paucity of parenchyma in conifer woods may relate to the tendency for conductive rates, if slow in tracheids, to demand less parenchyma for conductive requirements. Axial parenchyma is rather conspicuous in the two riparian conifers of New Caledonia, *Podocarpus minor* (plate 3-B) and *Dacrydium guillauminii*. In those two species, parenchyma abundance may relate to oxygenation problems of a plant with roots and trunk base under water.

Even with its torus mechanism, the coniferous bordered

pit is not a perfect valve. As Bailey (1958) indicates, elliptical pit apertures ("piceoid pits"), which cannot be closed completely by a circular torus, occur widely in conifers, especially in latewood and reaction wood. Moreover, pits in conifer latewood have thicker pit membranes and resist aspiration (closure of the pit by displacement of the torus) according to Bailey (1958) and Siau (1971).

Significance of helices (spirals) in secondary xylem tracheids.— Tracheids in some genera of conifers and taxads bear, inside the pitted secondary wall, helical bands (e.g., Taxaceae) or transverse bands (*Callitris glauca*), like the "tertiary helical thickenings" of some angiosperm vessels. The systematic and ecological distribution of these makes their interpretation quite difficult. For example, in *Picea, Pseudotsuga, Tsuga,* and Taxaceae, spirals are characteristically present; they are characteristically absent in other groups, notably Araucariaceae. Their occurrence in some Cupressaceae, but not in others (see Greguss, 1955) forms no immediately apparent pattern. As with helices in vessels of angiosperm woods, the proportion of conifer woods with helical thickenings in tracheids appears to increase with increasing latitude, but the exceptions are perhaps more notable than any percentages one could calculate. Helices would theoretically offer a mechanical strengthening ideal for countering collapse under conditions of high negative pressure in xylem, but we have no measurements to validate this yet. For the conifers where water tensions have been measured, no correlation with presence of helices appears evident. Spirals might enhance flexibility of wood, and tend to resist deformation of tracheids under conditions of high wind thrust. However, helices may, in fact, have little significance in conifers. In only a few conifers and taxads are they pronounced. Until further significance can be demonstrated, we can say that some conifers characteristically terminate increase in wall thickness of tracheids with formation of helices, whereas in others, the wall is smooth (or minutely warty) when wall formation has been completed.

THE ROLE OF MYCORRHIZAE IN CONIFERS

The most recent relevant survey of mycorrhizae, that of Marks and Kozlowski (1973) reveals that several functions can be attributed to mycorrhizae in woody plants. As Meyer's (1973) article indicates, mycorrhizae aid in supply of nutrients, particularly nitrogen, in areas such as boreal forest where, "The mineralization of organic nitrogen is hampered." Bowen's (1973) paper suggests greater drought resistance in mycorrhizal plants, with the mycorrhiza attenuating water supply as soil dries out. A corollary of this might be that mycorrhizae feed less than maximal water supply into roots when soil is moist, because a large root-hair system might be a more efficient means of rapid absorption than the intermediary device a mycorrhizal association offers. If so, mycorrhizae would supply water to a tree at a steadier rate, with fewer fluctuations, than would a root-hair system. Conifers would be excellent candidates to take advantage of such a system.

Conifers that grow in consistently mesic sites—Taxodiaceae, for example—would not be strongly advantaged by mycorrhizae. Meyer (1973) notes that Taxodiaceae are non-mycorrhizal and may now be restricted for this reason. However, he feels *Pinus*, which is ectomycorrhizal, may be advantaged and therefore be evolutionarily more aggressive. The advantage would be a capability of entering mineral-poor sites, and habitats with more sudden drops in soil-water availability. *Pinus* does appear to have entered such sites. The largest genus of conifers, *Podocarpus*, is non-mycorrhizal (Meyer, 1973). The radiation of *Podocarpus* is largely in the tropics, particularly in tropical uplands, where mycorrhizal forests are not as common as in the boreal regions. *Podocarpus* can be said to occupy consistently mesic sites in comparison with *Pinus*.

᎒᠍8.

Monocotyledons;
Nymphaeales

The papers by Cheadle (1942, 1943a, 1943b, and subsequent papers) as well as the various cited volumes of the *Anatomy of the Monocotyledons* have given us much information that can be interpreted in terms of origin and specialization of vessels in monocotyledons. Because all of the volumes of the *Anatomy of the Monocotyledons* have not yet appeared and because I wished to include all families, I have added a few new data. Vernon Cheadle (personal communication) informed me that *Aponogeton* lacks vessels in all portions. I prepared sections of *A. distachyus* and confirmed this finding. In order to supply information on Triuridaceae, I sectioned material of *Andruris sciaphila* I collected on Mt. Salawaket, New Guinea. This specimen has only tracheids in roots, stems, and inflorescence axes (end walls of tracheids in stems have wide scalariform pits, however; see plate 3-E). There appear to be no published data on tracheary tissue of Burmanniaceae. I sectioned material on two leafy species: *Burmannia disticha* (Australia) and *B. longifolia* (New Guinea). Both proved to have scalariform perforation plates on vessels, and such vessels were present in roots, stems, and leaves. Unfortunately no data are as yet available on tracheary elements of the heterotrophic (mycoparasitic) species of Burmanniaceae.

Cheadle's more recent papers (1968, 1969, 1970; Cheadle and Kokasai, 1971, 1972) have featured an accurate and sophisticated way of summarizing xylem for tribes and families. Cheadle rates vessel elements in each xylem sample from 1 (scalariform plates only) to 5 (simple perforation plates only), with 0 representing tracheids only. For each tribe and family, Cheadle's recent papers give a figure on this scale for early metaxylem, late metaxylem, and all xylem of roots, stems, inflorescence axes, and leaves, respectively. Such data deserve attention in their original form, and the representation given here as figure 8 is an oversimplification (as any graphic representation of this material is bound to be). The graph of figure 8 omits all families that lack vessels; these are discussed later.

In order to interpret figure 8, the reader must understand the conventions that were used. The range of vessel type and organographic occurrence is given for each family. Liliaceae has been used in the broad sense, and includes all segregate families commonly in use (Agavaceae, Alstroemeriaceae, Amaryllidaceae, Haemodoraceae, Petermanniaceae, Petrosaviaceae, Ruscaceae, Smilacaceae, Tecophilaeaceae, Trilliaceae, and Xanthorrhoeaceae). Musaceae as used here likewise includes the segregate families Heliconiaceae, Lowiaceae, and Strelitziaceae. Musaceae and Zingiberaceae have been enclosed within a single ellipse only because they show a very similar range of vessel types and occurrence, and space was thereby saved. Considerable overlap of families occurs in the upper left rectangle of figure 8. In some cases, genera have been cited on figure 8 (e.g., in the families Arecaceae and Liliaceae) to indicate which genera or subfamilies represent extreme vessel patterns in families with a range of types. The position of families relative to each other within each of the rectangles has no significance. For example, no distinction in vessel types was intended by placing Potamogetonaceae and Araceae at different ends of the upper left rectangle. No phylogenetic relationships should be interpreted from the placements, although some of the adjacent families probably are close (e.g., Sparganiaceae and Typhaceae). Exemplary of difficulties in representation are the following examples: Vessel elements have scalariform plates in roots, stems, and leaves of

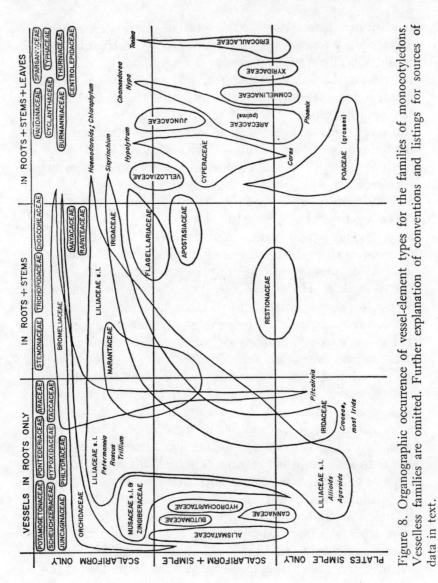

Figure 8. Organographic occurrence of vessel-element types for the families of monocotyledons. Vesselless families are omitted. Further explanation of conventions and listings for sources of data in text.

the Burmanniaceae studied, so that family rightly falls in the upper right rectangle. However, *Sisyrinchium*, also in this rectangle, has scalariform plates on vessel elements of leaves, but simple ones in the root. This is only a problem in the families shown to range across more than one rectangle. The chart obviously does not attempt to represent size of families. A crude attempt has been made, however, to indicate what proportion of a family for which xylem data are known has a particular pattern. The pointed tips of ellipses that barely enter a rectangle indicate that a minority of a family represents that condition.

A few curious distributions could not be shown. For example, *Phytelephas* (Arecaceae) has vessels in leaves and roots but not in stems (Tomlinson, 1961). The same is true in *Cordyline* and *Dracaena* of the Liliaceae (Cheadle, 1943*b*). Dioscoreaceae have vessels in roots and stems, but only in the petiolar portions of leaves (Ayensu, 1972). Inflorescence axes have been omitted from figure 8 because they often tend to conform to the pattern of stems, but more as a practical consideration, because information on inflorescence axes is not uniformly present in the volumes of *Anatomy of the Monocotyledons*.

AN ECOLOGICAL THEORY OF MONOCOTYLEDON
XYLEM SPECIALIZATION

Cheadle claimed that in monocotyledons, vessels originated first in roots, and have phylogenetically progressed into stems, inflorescence axes, and leaves, in that order. This progression, Cheadle avers, is paralleled by a shift from vessels with scalariform perforation plates to those with simple plates. The former type of vessel elements tend to be longer, the latter shorter. Cheadle's reasoning is presented in his 1942 paper. His general phylogenetic conclusions seem justified, and the question that needs particularly to be answered at present is why these trends have occurred. Why have particular phylads retained primitive conditions while others established specialized conditions?

In answering these questions, one must keep in mind how
monocotyledons differ from dicotyledons. Most dicotyledons
have taproots and lateral roots capable of secondary growth,
and in dicotyledons vessels are hypothesized to have originated
in roots and stems simultaneously. The adventitious nature of
roots in monocotyledons is significant, because each such root
has a finite duration, whereas the stem is relatively permanent in comparison. I would like to stress this because each
root of a non-aquatic monocotyledon may be expected to absorb water rather rapidly at first, then wither as water stress
occurs (except where roots have a duration of more than one
season, as in arborescent forms). Whereas the main root of a
woody dicotyledon does not die with the onset of drought,
roots of a non-arborescent monocotyledon often do. The most
persuasive consideration, however, is not that of water stress
but of sudden availability of water. The "seasonality" cited in
figure 9 should be interpreted not so much in terms of dryness
as in terms of a brief wet season during which large volumes
of water per unit time must be conveyed to the shoot system.
In other words, mesic habitats are mesic for indefinite periods,
whereas seasonal habitats are mesic for a limited portion of
the year.

The origin of vessels in roots of monocotyledons would not
be likely, however, if conductive rates were the same in all
parts of a plant equally, for if so, vessels should be present in
the same state of specialization throughout a plant. Thus,
explanation of differential distribution of vessels within the
plant body of a monocotyledon must take the form of explaining how differential conductive characteristics can occur
in roots, stems, inflorescence axes, and leaves, respectively,
within a single plant.

Let us take first the instance of those monocotyledons that
have developed vessels in roots only. For example, in allioid

and asphodeloid lilies, which have bulbs, or in *Iris*, which has rhizomes, the succulent perennating organ lacks vessels. Not only do the roots in these groups have vessels, these vessels have simple perforation plates as well. During the brief wet season typical of species in these genera, water must be conveyed rapidly to the shoot system. Succulence of the bulb or rhizome presumably connotes slow conductive rate within the shoot system.

The comparisons of species in Cheadle's (1969) paper dealing with amaryllids are informative. The species of the tribe Allieae all have simple perforation plates in roots. A species of this tribe figured by Cheadle, *Muilla maritima*, leafs out and flowers during the brief moist winter of California's Coast Ranges. One can contrast this with *Agapanthus*, in which perforation plates in roots are scalariform. *Agapanthus* is typically a plant of marshy streams in Africa, sites that provide prolonged water availability during the year. Cheadle's comment (e.g., 1969) that monocotyledon groups with more primitive xylem cannot have been derived from stocks with specialized vessel types is entirely justifiable. In this case, we must hypothesize that the monocotyledons with more primitive xylem configurations have had an unbroken history of occupation of more mesic habitats (unless reversibility can occur, and very likely only a moderate degree of reversibility is possible, judging from the totality of data).

Certainly some groups with more specialized xylem can, during phylesis, invade more mesic habitats and are not disadvantaged there. However, relatively few families and genera of monocotyledons appear to have done so. This suggests that monocotyledons are not strongly disadvantaged by primitive xylem characteristics compared to dicotyledons. Judging by the perforation plate type of vessel elements, a much higher proportion of dicotyledons have specialized types of conductive tissue. Measurement of conductive rates and transpiration rates is very much needed, with species selected in terms of differential types, as shown in figure 8. One would guess, for example, that rates of flow in stems of Pandanaceae are not rapid compared to those in grasses.

The data of Bierhorst and Zamora (1965) on primary xylem of dicotyledons show that in general primary xylem has more primitive tracheary elements than does secondary xylem. Their data suggest that primary xylem of dicotyledons is, in fact, relatively comparable in level of specialization to that of monocotyledons at large.

As noted for gymnosperms, tracheids have a potential advantage in localizing any cavitation within individual cells, whereas an air embolism can disable an entire vertical series of vessel elements. Thus, tracheids in monocotyledons may be retained as long as they are not positively disadvantageous in terms of conductive rates. As we will see, probably all monocotyledons are capable of repair of air embolisms in vessels by means of root pressure, a repair which could conceivably take place at any time when transpiration rate is markedly reduced, as at night.

Water loss from bulbous or rhizomatous monocotyledons is apparently not great. Leaves tend to wither as soon as soil water becomes low; otherwise, leaves would begin to withdraw water from the bulb or rhizome more rapidly than it could be replaced. Exceptions occur where leaves are minute, or scale-like (*Asparagus*), or markedly succulent. Such adaptations as these would tend to reduce transpiration to a steady and low rate. This situation differs from that of the ferns, where leaves must be broader, and where relatively few leaves seem adapted for lowered transpiration (although the data by Gates, 1968, for high diffusive resistance in *Pteridium aquilinum* suggest low transpiration rates without microphylly or succulence). If transpiration is relatively high in ferns, this would explain the fact that pits in overlap areas of tracheids tend to be larger than those in stems, but smaller than those in roots.

A monocotyledonous leaf succulent or stem succulent provides a strong differential among organs. In fact, a leaf succulent such as *Aloë* or *Agave* tends to have a succulent stem as well, but roots that function only during the wet season. Both of those genera have simple perforation plates in root vessel elements, but no vessels elsewhere in the plant. We can note that vessels are highly specialized in roots of most Iridaceae (and by "specialized" I would connote "adapted to rapid water conduction"). The tribe Aristeae of Iridaceae has notably primitive vessel elements in roots, according to Cheadle (1963). Aristeae are chiefly montane South African plants, and grow in localities more mesic than one might suppose, either by virtue of frequent condensation from clouds or availability of underground water. These localities also host dicotyledons with more primitive xylem, such as Bruniaceae, which have scalariform perforation plates in wood. Aristeae, incidentally, do not have succulent rhizomes. *Sisyrinchium* is exceptional among irids in that vessels occur in leaves. This seems correlated with the fact that *Sisyrinchium*, which occurs in regions with dry summers, does not have a succulent rhizome with early withering leaves. This is also true of the only member of Liliaceae in the strict sense (i.e., Lilioideae here) studied by Cheadle in his 1942 paper, *Chlorophytum elatum*. I have deliberately withheld water from plants of *Chlorophytum elatum* cultivated in Claremont, but leaves did not wither during hot summer months. This capability may be related to low transpiration rate of leaves and succulent tuberous roots of this species. These factors would mean that conductive rates would be not strongly different throughout the plant.

Monocotyledons With Secondary Thickening

The exceptional nature of *Cordyline* and *Dracaena* is worthy of mention. These genera have vessels in leaves, but only tracheids in stems. This may be correlated with their relatively broad, nonsucculent leaves that might transpire fairly rapidly during periods of low humidity and intense insolation.

This activity would not affect stems so much (because the stems can be considered succulent), and thus a differential between leaves and stems could be said to exist. Cheadle (1943b) suggested that the abundant volume of tracheids in these stems compensates for lack of vessels in stems. That may be true. However, I note that all the agavoid genera studied by Cheadle, other than *Cordyline* and *Dracaena*, have succulent (*Agave*) or xeromorphic (*Nolina*, *Yucca*) leaves.

There may also be a correlation between the production of secondary bundles and the nature of the elements in them on an ontogenetic basis. If vessels were added by the "mono-cotyledonous cambium," establishment of a continuity between them and the vessels of the upper primary stem would be morphogenetically a virtual impossibility. Addition of tracheids as a conductive tissue also has the advantage that air embolisms, if they occur, would be localized within individual tracheids as noted above for conifers. If strong negative tensions make xylem subject to such cavitations or if xylem of "woody" monocotyledons becomes deactivated with age, addition of tracheids would be valuable. Tracheids certainly would suffice for conductive capacity if transpiration and conduction rates were slow.

One may be surprised that some trees of desert or very dry regions—the arborescent species of *Yucca* and its allies—do not have vessels in stems and leaves. However, there is evidence that leaves, and therefore stems, have low conductive rates. In at least some stem succulents, transpiration occurs at night, a time at which humidity is higher, insolation absent, and transpiration rates therefore minimal. This requires a nocturnal carboxylation cycle. Thus far, this type of photosynthesis has been demonstrated for *Agave* among monocotyledons, and *Aeonium* and *Bryophyllum* among dicotyledons (Neales, Patterson and Hartney, 1968). Roots of *Yucca* and *Agave*, however, must be able to convey volumes of water rapidly to the shoot system during months of water availability, and presence of vessels in roots is therefore understandable.

One may note at this point that "monocotyledonous secondary growth" is restricted to the woody liliaceous genera and to a few members of the tribe Aristeae of Iridaceae (Tomlinson and Zimmermann, 1969). Of the 14 genera listed by them as having this secondary activity, only 2, the xanthorrhoeoids *Kingia* and *Xanthorrhoea*, do not ordinarily branch. Secondary activity is relatively limited in stems of *Kingia* and *Xanthorrhoea*. In any case, the increase in leaf-bearing branches is probably correlated with increase in the number of bundles by the meristematic activity in the 12 remaining genera.

Large Monocotyledons Without Secondary Bundles

Very few palms branch (except from the base, in which case the offshoots develop their own root systems). This lack of secondary activity in palms may be related to a lack of increase in the number of leaves. Arborescent Musaceae (*Strelitzia nicolai, Ravenala madagascariensis*) typically do not branch except from the base. To be sure, in Pandanaceae, the ever-widening stem is supported by prop roots of ever-increasing diameter, originating at progressively higher levels on the stem. These thicker upper prop roots provide sufficient water almost directly to the crown, and circumvent the need for secondary growth. In palms, a few—*Iriartea, Socratea,* and *Verschaffeltia,* for example—do have prop roots. Roots in palms do not increase so markedly in diameter with height and age of the plant. When the large prop roots no longer form or are able to reach the ground in *Pandanus,* the plant declines, in fact. Taller species of *Pandanus* tend to be less branched and are often unbranched. This would correlate with the inability to form new contacts with the ground by prop roots, so that a vegetative pattern more like that of palms results.

Secondary activity would also be of no physiological value in *Freycinetia* (Pandanaceae), because continually innovated adventitious roots of this lianoid plant supply each shoot (if not, that shoot fails). In its habit, *Freycinetia* mimics the climbing aroids, which also lack secondary addition of bundles.

Marsh Plants

In the marsh and aquatic plants of the upper left rectangle of figures 8 and 9, the fact that roots are inundated or in very wet soil, yet the roots have vessels, many seem ironic. However, the sites in which they grow do fluctuate in substrate water availability. *Triglochin* (Juncaginaceae) grows on estuarine and other flats that can dry at times or seasonally. This drying is usually not so extreme as that of the habitats of the monocotyledons with more specialized vessels. If drying does occur, the plant may die entirely except for a storage organ. In *Potamogeton*, for example, vessels occur only in roots and these have scalariform perforation plates. Highly efficient vessel elements would be of no value in situations like this; primitive vessel elements would cope with the minor fluctuations in the habitat of these monocotyledons.

Butomaceae, Alismataceae, and Hydrocharitaceae have more specialized vessels in roots. These plants may be hypothesized to persist in conditions with greater fluctuation of water availability than those of the families of the upper left rectangle of figure 8, or to have bulbs or bulb-like structures (*Alisma, Butomus*) that retard cessation of activity during drying of the habitat. Some species of *Damasonium* and *Echinodorus* (Alismataceae) are annuals, and therefore roots begin growth in marshy conditions but terminate in dry soil —a condition reminiscent of *Marsilea*.

Forest Understory Plants

The occurrence of scalariform perforation plates in vessels, and vessels in roots only in terrestrial forest genera of the upper left rectangle is readily understandable in terms of the slow progression of lowered water availability. In fact, if we notice the leaf morphology of these, we notice that most are quite broad-leaved (e.g., *Trillium, Ruscus, Arisaema, Tacca*),

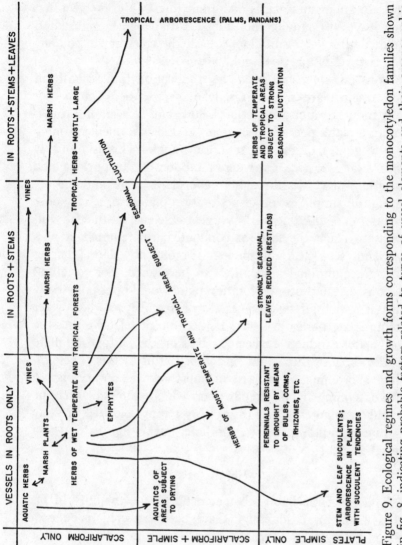

Figure 9. Ecological regimes and growth forms corresponding to the monocotyledon families shown in fig. 8, indicating probable factors related to types of vessel elements and their organographic occurrence.

and are understory plants of humid forest. The broad, often thin leaves of many of these suggest that high humidity and shade diminish transpiration, so that the efficiency of conduction that vessels provide would be unnecessary.

Thus far, there has been the implication that scalariform perforation plates are not disadvantageous where volume per unit time of water conduction is low; and so scalariform perforation plates persist (or in stems and leaves, tracheids only) in the groups of the upper left rectangle. This is probably true as far as it goes, but where the potential for phylogenetic extension of vessels throughout the plant exists, and the potential for simplification of perforation plates exists, why does this not happen? One might say that efficiency of the end wall tends to reach the level of conductiveness required, a phenomenon seen clearly in phloem of monocotyledons (chapter 12). An additional possibility can be offered. For structural reasons, a scalariform perforation plate would offer potentially greater structural rigidity (e.g., against collapse when water columns are under tension). If, because of selective pressure for greater conductive efficiency, bars on the perforation plate tend to be lost, there must be compensating factors. These include shortening, an increase in wall thickness, and change to alternate circular pits, trends upon which comments may be found in chapter 11. Thus, a fairly strong selective pressure can be hypothesized to alter not one, but several features of a vessel element.

Epiphytes

The epiphytic habit, characteristic of the majority of orchids and bromeliads, might have been expected to be associated with greater specialization of vessels in roots. We must concede that published knowledge of the occurrence of vessels in Orchidaceae is, as yet, based on a small number of genera, and when more knowledge becomes available, ecological correlations may become more obvious. For example, our knowledge of vessel types and the organographic distribution of vessels in Bromeliaceae is wider now that we have Tomlinson's (1969) account than it was on the basis of Cheadle's (1942) survey. In any case, several factors might explain the lack of specialization of vessels in roots of the epiphytes. Water stress on roots may not be great or prolonged, and it is not progressive in the way that it would be for a plant growing in soil that becomes drier at the end of a growing season. Any habitat that is favorable for epiphytes has only occasional lack of moisture, not prolonged drought, and relatively high humidity. The roots of epiphytic orchids are so constructed that the root is, in essence, succulent. The velamen, an outer cortex of dead cells, provides storage of water during intervals between rainfall or condensation from clouds. The velamen thus attenuates water availability considerably. The structure of velamen shows a series of thickened tangential walls that deter evaporation of water. The endodermis and innermost cortical layers of roots of orchids are constructed as a sheath, with both mechanical and water-canalization capabilities. The endodermis is thin-walled opposite xylem poles in absorptive portions of roots, but older portions show a totally fibrous endodermis. The cortex of older portions of an orchid root can be dead, in fact, while conductive tissues are still functioning. The "tank bromeliads" (species that collect water in leaf bases) can rely more on leaves than on roots for water absorption. Roots in these species become more a means of securing the plant to its substrate than a water-absorption system. In *Pitcairnia*, a terrestrial bromeliad, reliance on soil moisture seems obviously related to simple perforation plates in vessel elements of roots.

Tropical Mesophytes; Typhales; Centrolepidaceae

The plant families and genera located in the upper right rectangle of figure 8 can be divided into several classes, as suggested by the interpretive diagram of figure 9. Among these are tropical herbs, including some with fairly large plant size (a fact no doubt related to greater water availability and lack of cessation to the growing season). These tropical forms grow in moist soil: for example, Cyclanthaceae, *Chamaedorea* (Arecaceae), and some Flagellariaceae are typically understory elements in tropical forests. *Nypa* is distinctive among palms in its riparian habitat. These and other genera of the humid tropics have foliage immersed in a moist atmosphere; Pandanaceae and Burmanniaceae have been mentioned earlier. The fact that the scalariform perforation plates occur in vessels of leaves as well as in those of roots and stems is comprehensible: there is a rough equality between the expanse of root and leaf surface and the relative constancy in amount of soil and air moisture.

The occurrence of two temperate families of marsh plants, Typhaceae and Sparganiaceae, in the upper right rectangle of figure 8 may seem unexpected. However, one may note that Typhaceae and Sparganiaceae tend to be large marsh plants, with long persistent leaves emerging from the water, that would experience greater transpiration than plants with floating leaves (such as Potamogetonaceae). Because of condensation in leaf form, however, their rate of transpiration is moderate. *Thurnia* (Thurniaceae) may, although tropical, be likened to *Typha* or *Sparganium* in its habit and habitat. The perennial cushion plants of Centrolepidaceae grow on very moist substrates, so that one could make a case for aptness of primitive vessels in leaves of that family.

Families With High Xylem Specialization

Increasing seasonality (figure 9) has probably been basic to specialization of vessels in such families as Cyperaceae, Juncaceae, Eriocaulaceae, Xyridaceae, and Arecaceae. In Cyperaceae, few genera have scalariform perforation plates throughout the plant. Those that do, like *Hypolytrum*, are tropical forest elements, occupying habitats like those of Cyclanthaceae or leafy Burmanniaceae. *Carex* and other Cyperaceae with more specialized vessels occur in temperate environments with greater extremes of dryness in soil and air. The palms with the most specialized vessels, such as *Phoenix*, tend to be those that range out of tropical forests into less tropical, drier, savannah-like habitats. Some Flagellariaceae and Velloziaceae can occupy open tropical and subtropical habitats, and their vessel conditions do reflect a moist savannah habitat.

Eriocaulaceae, Xyridaceae and Commelinaceae are tropical and subtropical families, and one might not expect them to have xylem with vessel elements so highly specialized, like those of grasses. However, they can occupy areas that are seasonally extremely dry; they do not lose their leaves during these periods. That *Tonina fluviatilis* should have xylem of lesser conductive efficiency (figure 8) correlates with its riparian habitat.

The possibility definitely does exist that monocotyledons adapted to water stress can enter more mesomorphic conditions. This possibility, mentioned by Kokasai, Moseley and Cheadle (1970) can be hypothesized in certain instances with considerable certainty. For example, the species of *Panicum* in Hawaiian bogs, such as *P. isachnoides*, occupy areas with more than 400 inches of rain per year and could hardly be imagined to have stemmed from ancestors accustomed to conditions wetter than that.

The situation in Poaceae as a whole, however, is not clear. Soderstrom and Calderon (1971) claim that among grasses, species of the tropical forests may have more numerous features which could be considered primitive for the family. Assuming that this is true, either grasses originated in such habitats or some phylads of grasses were early entrants into tropical forest and have remained little changed since. The rapidity with which shoots elongate and perhaps also the basal meristem mode of growth (a parallel may be noted to *Equisetum*, which has these features and has vessels) could be related to a high degree of xylem specialization in grasses. In any case, grasses seem preadapted by virtue of their conductive tissue for habitats of strong fluctuation in water availability.

The fact that some palms have vessels as specialized as those of grasses is not difficult to understand. Their large leaves, often reaching the forest canopy if the species is, in fact, not characteristic of open habitats, are subject to transpiration of large volumes of water if humidity is low and temperature and insolation are high. The fact that palm stems must conduct large volumes of water rapidly through a limited number of vessels, together with the fact that no additional bundles are formed by secondary activity, give a positive selective value to simple perforation plates in palm stems. Further comments on problems of water-conduction in palms are given below.

Restionaceae (figs. 8, 9) have a level of vessel specialization comparable to that of Cyperaceae or Poaceae, except for lack of vessels in leaves. This can be clearly attributed to reduction in leaf surface and to the non-photosynthetic nature of scale leaves of Restionaceae. Only the first few leaves of a seedling are photosynthetic, and these are very small in Restionaceae. The remainder of the plant bears only scarious leaves. Probably only a situation like this would lead to presence of highly specialized vessels in roots and stems, but lack of vessels in leaves. Restionaceae occupy mediterranean-type habitats, so that a high degree of vessel specialization would be expected.

Vesselless Aquatic Monocotyledons

Among vesselless families of monocotyledons are the families Aponogetonaceae, Lemnaceae, Najadaceae, Ruppiaceae, Zannichelliaceae, and Zosteraceae. Absence of vessels in Aponogetonaceae might not be expected, because this family has plant size approximately like that of Potamogetonaceae. Assuming that all Aponogetonaceae, when studied, will prove to lack vessels, this is still understandable. In *Aponogeton distachyus*, the wide pit areas on tracheids of roots and stems serve as maximal areas across which water transfer can occur as could be possible short of origin of actual vessels. In this connection, data on width of pits in overlap areas in fern tracheids (table 7) are relevant. Cheadle (1943*a*) reported end walls (specialized overlap areas) on tracheids of *Clintonia* (Liliaceae) and *Acorus* (Araceae) stems. Such tracheids also occur in Triuridaceae (plate 3-E). An interesting subject for investigation is whether, as the fern data suggest, increasing width of pits in tracheid overlap areas foreshadows the origin of vessels. For example, do the families in which vessels occur only in roots (Pontederiaceae, Hypoxidaceae, etc.) have wider pits in the tracheid overlap areas in stems than in leaves?

The vesselless aquatic families do seem to lack vessels because of reduction in vascular tissue, as Cheadle (1942) hypothesizes. The families listed (except for Lemnaceae) can be called "helobioid" families; they would thus be related to Potamogetonaceae, among vessel-bearing monocotyledons. On account of this relationship, I would guess that the vesselless families have been derived from phylads, in which vessels, if present, occurred in roots only. Broad pit membrane areas between tracheids adjacent in a file in the vesselless families offer sufficient conductivity for the slow rate of water transport in these plants. An intriguing question posed by the vesselless monocotyledons is to what degree can reversibility in vessel evolution occur? If tracheids are present in protoxylem, and vessels only in the late metaxylem of a given monocotyledon, changes in growth form (such as adaptation to an aquatic habitat) that are accompanied by reduction in vascular tissue (i.e., formation of less metaxylem as it were) could produce wholly tracheidal bundles.

Mycoparasites

Another category of vesselless monocotyledons must be recognized: mycoparasites. Mycoparasites are flowering plants, formerly termed saprophytes, in which the plant is achlorophyllous and all nutrition and water is obtained via a mycorhizal association. Triuridaceae is a family of mycoparasites. Dr. Margaret Stant (personal communication) has reported that in three genera of Triuridaceae, vessels are lacking throughout the plant. Tracheids in stems have specialized end walls, a feature I confirmed for *Andruris sciaphila* (plate 3-E).

Petrosavia (Liliaceae; sometimes considered as a separate family, Petrosaviaceae) is also a mycoparasite. It lacks vessels in roots as well as in the remainder of the plant (Margaret Stant, personal communication). *Triuridaceae* and *Petrosavia* are all plants of the understory of wet tropical forests. Under these conditions, water is constantly available. Because the mycoparasites grow slowly, water is probably fed to the parasite by the mycorrhiza at a slow but steady rate, perhaps even with little diurnal flunctuation. Under these conditions, a vascular system for rapid water conduction would be incongruous. Transpiration is minimal in plants with the habit and habitat preferences of Triuridaceae and *Petrosavia*, so that demands on a conductive system should surely be minimal. Vascular tissue is relatively poorly developed in these plants. One can hypothesize that in autotrophic ancestors, vessels were present in roots, but that vessels in roots have been lost by reduction. The phylogenetic position of these families suggests that ancestors may never have had vessels in stems.

In *Andruris sciaphila*, end walls show scalariform pits, but lateral walls of tracheids have only a few pits (plate 3-E). This may be considered a mechanical enhancement of the tracheid lateral wall simultaneous with a conductive enhancement of the end wall, similar to the tendencies in fern tracheids (chapter 2). Tracheids in *Andruris sciaphila* are suggestive of mechanical strength because they are associated with sclerenchyma in the stem.

WATER TRANSPORT IN MONOCOTYLEDONS

Ray (1972, p. 85) has stated:
"An interesting problem is presented, however, by palm trees, which during their life span do not add any new vascular elements whatever to the stem except at the top where the tree is growing in height. Palm trees often exceed 33 feet in

height and must therefore sustain continuous tensions in their xylem, so it is hard to see how they could avoid succumbing to air embolisms unless they possessed some kind of active mechanism for removing air bubbles from the xylem sap. No such mechanism is known, however."

Such a mechanism does seem to have been discovered, however. Davis (1961) reports exceptionally high root pressures in palms, effective to heights above 10 m. Root pressure has generally been regarded in recent years as a secondary phenomenon in the total picture of water uptake in vascular plants. Perhaps it often is. If, however, root pressure can account for water translocation in palms, it certainly could be a factor of prime significance in the picture of water translocation of other monocotyledons. An explanation of this sort is put forth by Epstein (1972) when he says, "it is likely that once the rate of transpiration falls to a low value, as at night, these ruptures in the liquid columns in the xylem are eliminated, and that root pressure furnishes the necessary force."

We can imagine that development of negative tensions in leaf xylem of monocotyledons, as in other plants, does account for water transport to a large extent. However, breakage of water columns could be frequent, especially in plants with vessels throughout the plant. Thus, in a monocotyledon such as a grass, the xylem at the end of a day's transpiration might well contain air embolisms. At night these would be repaired by root pressure—a likely occurrence, if one notes the frequency with which guttation droplets can be seen on the tips of grass leaves early in the morning. In such a plant, a system of vessels throughout the plant would be advantageous, in that sap rising by root pressure could easily clear all embolisms from the entire shoot system in a short period of time. A corollary to this theory is that the monocotyledons with vessels throughout the plant are those in which air embolisms are more likely to develop, and that these plants can tolerate this. The lack of vessels in stems or leaves suggests a regime in which air embolisms rarely develop. To be sure, such cavitations, if they occur in tracheids rather than vessels, are limited to individual tracheids and do not spread the length of a bundle. Such potential cavitations presumably could be mended easily as tension in the conductive system is relaxed.

The presence of parenchyma around and in vascular bundles of monocotyledons has several possible functions. That it is functioning in some way other than as mere storage tissue is suggested by its distribution in palms. One might expect that palms, with their highly sclerotic stems, would have vessels surrounded by sclereids—a distribution that would lend strength to the bundles. Instead, vessels are always sheathed by parenchyma in palms (Tomlinson, 1961). Such parenchyma may well serve for lateral transport of water between bundles in a stem (particularly if a bundle is disabled), and undoubtedly serves to transfer photosynthates from the phloem into the ground tissue parenchyma, where it can be converted into starch.

VESSEL DIMENSIONS

Cheadle (1943a) notes that longer vessel elements are correlated with scalariform perforation plates in monocotyledons. This might be expected on the basis of similar patterns in dicotyledons, and reasons for the adaptive significance of vessel-element lengths are offered in chapters 10 and 11. Cheadle also reports exceptionally long vessel elements in *Asparagus* and *Dioscorea*. These two genera are exceptional in the degree of elongation of stems and the rapidity with which it occurs. These factors would tend to produce exceptionally elongate procambial cells. Vessels are short, however, in metaxylem of grasses, such as bamboos. Study of longitudinal sections of bamboo stems shows that transverse subdivision of metaxylem procambium occurs until just before elongation has ceased, so that this would be understandable. In bamboos, the great width of metaxylem vessels probably makes the short length

PLATES

Plate 1. Ferns. A. *Oleandra* sp., habit, Genting Highlands, Malaya; plant about 1 m. tall; note slender erect stems, leaves nearly sessile in pseudoverticils. B. *Matonia pectinata*, transection of central vascular strand of stem, showing parenchyma cells among tracheids. C. *Dicranopteris dichotoma*, transection of central portion of stele; tracheids are immature; parenchyma cells separating them contain tannins. D. *Osmunda cinnamomea*, transection of portion of vascular cylinder, showing separate strands of tracheids, parenchyma absent within strands. 1-B,C at magnification of scale shown below plate 8-E; 1-D at the scale shown above plate 5-A.

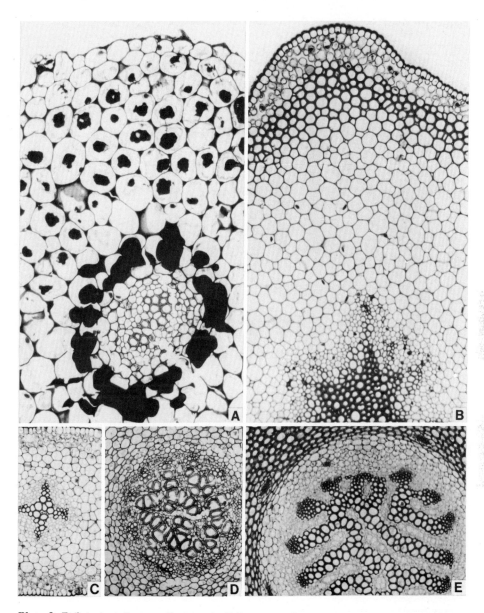

Plate 2. Psilotaceae, Lycopodiaceae. A. *Psilotum nudum*, transection of subaerial axis. B. *Psilotum nudum*, transection of aerial axis; tracheids, difficult to distinguish, are at periphery of the central strand of sclerenchyma. C. *Psilotum complanatum*, portion of transection of aerial axis, showing tetrarch xylem. D. *Lycopodium phyllanthum*, transection of stele and adjacent parenchyma. E. *Lycopodium sitchense*, transection of stele and adjacent sclerenchyma. All at magnification shown below plate 13-D.

Plate 3. Conifers, Monocotyledon. A. *Agathis australis* (Araucariaceae), transection wood; parenchyma cells may be seen adjacent to rays. B. *Podocarpus minor* (Podocarpaceae), transection wood from near the broad base of a small tree; parenchyma cells may be seen scattered among the tracheids. C. *Taxodium mucronatum* (Taxodiaceae), transection of wood, showing growth ring. D. *Microstrobos niphophilus* (Podocarpaceae), transection of wood, showing 6 growth rings, whereas A–C show latewood near bottoms of photographs. E. *Andruris sciaphila* (Triuridaceae), tracheid from radial section of stem; scalariform pitting of end wall visible. 3-A,B,C,D magnification at scale shown above plate 5-A; 3-E at the scale below plate 8-E.

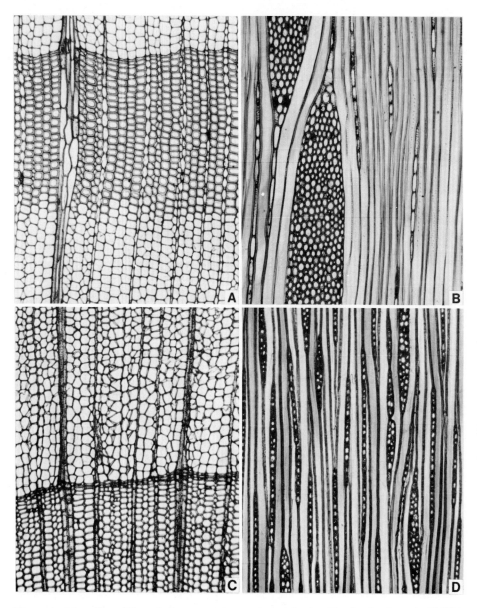

Plate 4. Vesselless Dicotyledons, sections of wood. A. *Trochodendron aralioides* (Trochodendraceae), transection, showing earlywood at top and bottom. B. *Trochodendron aralioides* (Trochodendraceae), tangential section; portions of two multiseriate rays at left. C. *Tetracentron sinense* (Tetracentraceae), transection; latewood below. D. *Tetracentron sinense* (Tetracentraceae), Tangential section through latewood. All at magnification scale shown above plate 5-A.

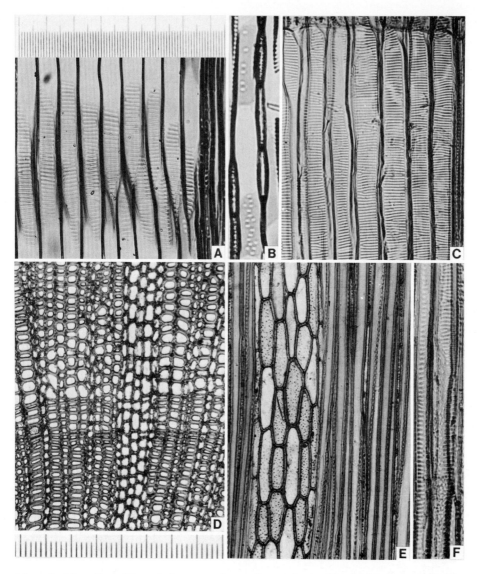

Plate 5. Vesselless Dicotyledons, sections of wood. A. *Trochodendron aralioides* (Trochodendraceae), radial section, showing overlap pitting on earlywood tracheids. B. *Trochodendron aralioides*, portion of tangential section, showing circular pits on tangential walls of tracheids. C. *Tetracentron sinense* (Tetracentraceae), radial section, showing pitting on earlywood tracheids. D. *Sarcandra glabra* (Chloranthaceae), transection; fluctuation in radial diameter of tracheids, suggesting a growth ring, is present. E. *Sarcandra glabra* (Chloranthaceae) tangential section, multiseriate ray at left. F. *Sarcandra glabra* (Chloranthaceae) portion of radial section, showing scalariform pitting on overlap areas of tracheids. Plate 5-A,B,C: magnification scale shown above A is a photograph of a stage micrometer with 10 μ divisions, enlarged at same scale as wood section; plate 5-D,E,F: scale shown below D.

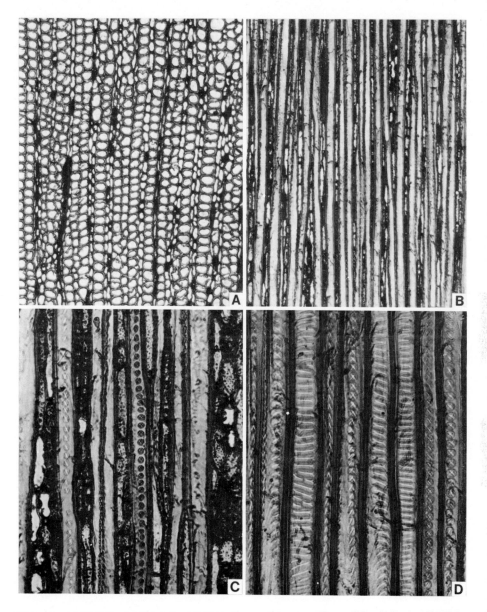

Plate 6. *Amborella trichopoda* (Amborellaceae), a vesselless dicotyledon. A. Transection; no growth rings occur. B. Tangential section shows abundance of uniseriate rays, narrowness of multiseriate rays. C. Tangential section, shows circular pits on tangential walls of tracheids. D. Radial section, showing scalariform pitting on overlap areas of tracheids. 6-A,B magnification at scale shown above plate 5-A; 6-C,D, scale shown below plate 8-E.

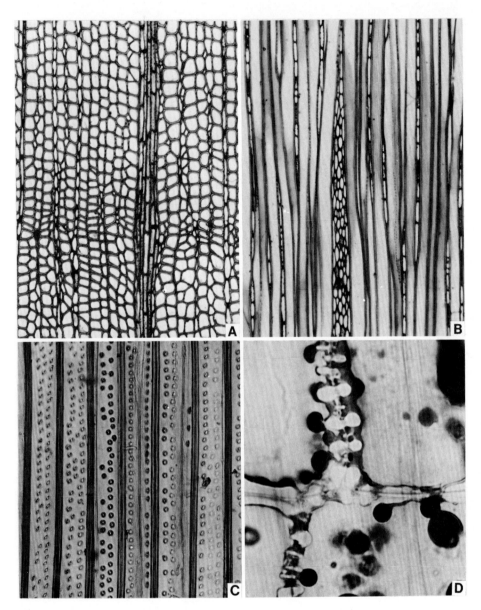

Plate 7. *Drimys winteri* (Winteraceae), a vesselless dicotyledon, sections of wood.
A. Transection; weakly marked growth ring apparent below center. B. Tangential
section, showing portion of one multiseriate ray. C. Radial section; overlap areas of
tracheids, showing circular bordered pits. D. Radial section; ray cell walls in sectional
view, showing some pits bordered, most pits outlined by dark-staining deposits.
7-A,B, magnification scale shown above plate 5-A. 7-C scale shown below plate 8-E.
7-D, four times magnification scale shown below plate 8-E.

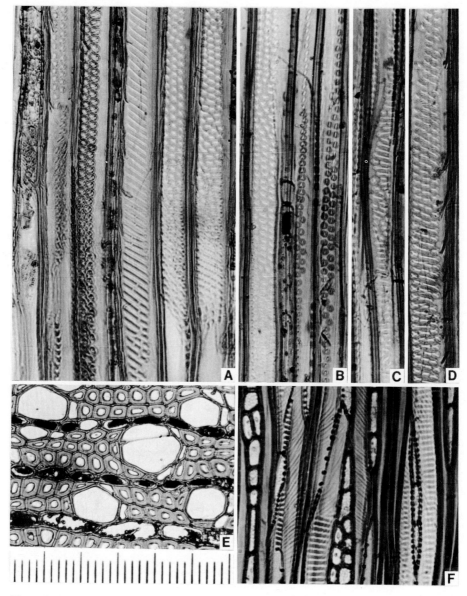

Plate. 8. Wood sections of Winteraceae, Illiciaceae, Magnoliaceae. A. Radial section, *Zygogynum* cf. *pomiferum*, showing scalariform pitting on overlap area of tracheid in center. B. Radial section, *Bubbia semecarpoides*; tracheids showing circular bordered pits on non-overlap areas. C. Radial section, *Bubbia semecarpoides*; tracheids showing scalariform pitting on overlap area. D. *Bubbia* cf. *sylvestris*, tracheid showing opposite and some elongate pits on overlap area of tracheid. E. *Illicium cambodianum* (Illiciaceae), transection; showing vessels angular in transection, thick-walled tracheids. F. *Michelia fuscata* (Magnoliaceae), tangential section; scalariform lateral wall pitting on vessels and helical thickenings can be seen. All magnified according to scale below 8-E, which shows 10 μ intervals of a stage micrometer that has been enlarged at the same scale as the sections.

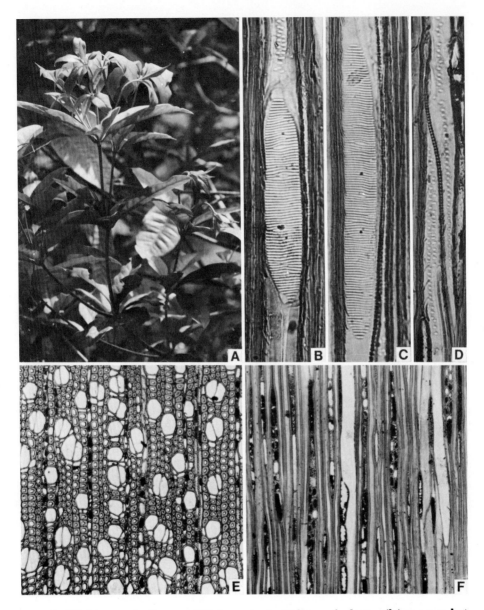

Plate 9. *Illicium cambodianum* (Illiciaceae). A. Foliage of plant wilting on a hot day, July 18, 1973 (Carlquist 4420), Maxwell Hill, Malaya. B and C. Perforation plates from radial sections. D. Wood section; pitting on tangential surfaces of vessels. E. Wood section; transection. F. Tangential section. Rays contain dark-staining deposits. 9-B,C, magnification scale shown below plate 8-E. 8-D; 9-D,E scale shown above plate 5-A.

Plate 10. *Pentaphragma horsfieldii* (Pentaphragmataceae). A. Habit of plant (Carlquist 4421), Maxwell Hill, Malaya. B. Transection; a few pith cells at bottom. C. Tangential section, showing rayless condition, septate living tracheids. D. Radial section; two vessel elements at left. E. Radial section; portion of perforation plate, showing bordered bars. 10-B,C,D magnification according to scale shown below plate 15-D; 10-E four times magnification of scale shown below plate 8-E.

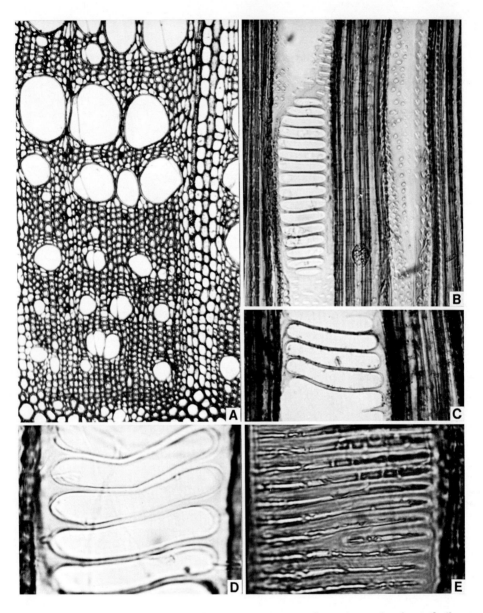

Plate 11. Austrobaileyaceae and Eupomatiaceae, wood sections. A. *Austrobaileya scandens*, transection. Vessels become wider from pith (bottom edge of photograph) outward. B. *Austrobaileya scandens*, radial sections. Perforation plate on narrow vessel; walls of vessel at right have alternate pits. C. *Austrobaileya scandens*; perforation plate of wide vessel element. D. *Austrobaileya scandens*; perforation plate at high magnification, borders show at bases of bars. E. *Eupomatia laurina*; perforation plate from radial section, wall material forms interconnections between bars. 11-A. Magnification according to scale shown above plate 5-A; 11-B,C scale shown below plate 8-E; 11-D,E, four times magnification of scale shown below plate 8-E.

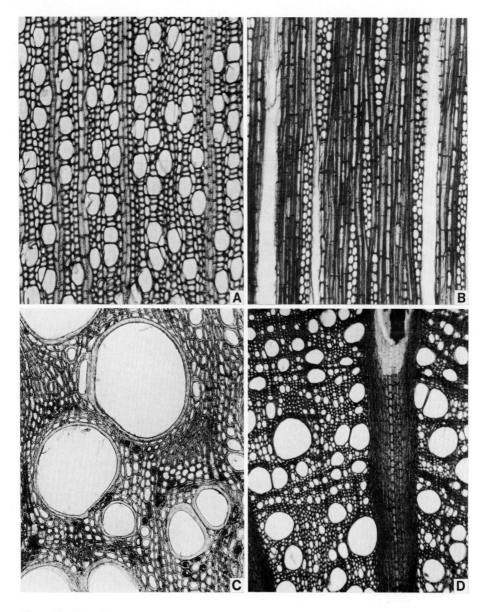

Plate 12. Wood sections of dicotyledons. A. *Leonia glycicarpa* (Violaceae); transection, axial parenchyma absent. B. *Leonia glycicarpa*; tangential section, septate fibers visible. C. *Doxantha unguis-cati* (Sapindaceae); transection of secondary xylem, vessels notably large, thick walled. D. *Akebia trifoliata* (Lardizabalaceae); transection of secondary xylem, ray at right. 12-A,B, magnification scale shown above plate 5-A; 12-C scale below plate 5-D; 12-D scale below plate 13-D.

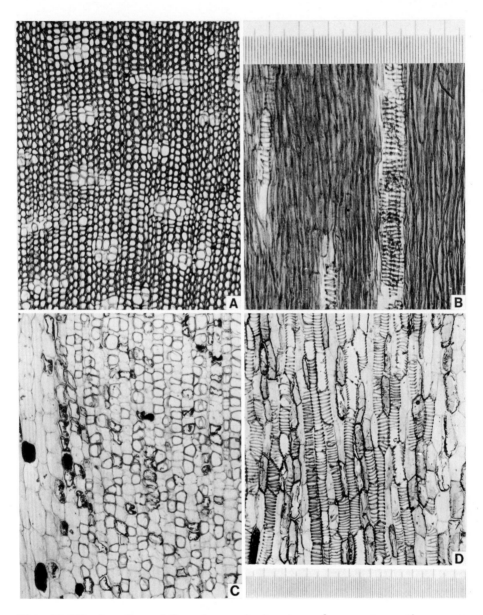

Plate 13. Wood sections of Crassulaceae. A. *Aeonium arboreum*; transection, group of vessels are visible in the background of libriform fibers. B. *Aeonium arboreum*; tangential section, rayless condition evident. C. *Crassula argentea*; transection, ray at left. D. *Crassula argentea*; tangential section, thickenings on vessel-element lateral walls simulate annular bands. 13-A,B, magnification scale (10 μ units) shown above B; 13-C,D, scale below D.

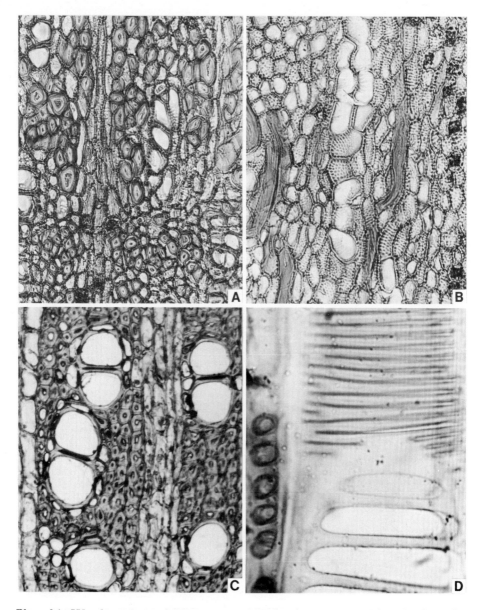

Plate 14. Wood sections of Viscaceae and Rhizophoraceae. A. *Phoradendron flavescens macrophylla*; transection, gelatinous fibers can be seen. B. *Phoradendron
flavescens macrophylla*; tangential section, vessel elements are about as long as they
are wide. C. *Ceriops tagel*; transection, vessels and tracheids are thick walled. D.
Ceriops tagel; vessel from radial section, three of six bars of perforation plate can be
seen (below), lateral wall pitting (above). 14-A,B, magnification scale shown below
plate 15-D. 14-C, scale shown below plate 8-E; 14-D, four times the magnification
of scale shown below plate 8-E.

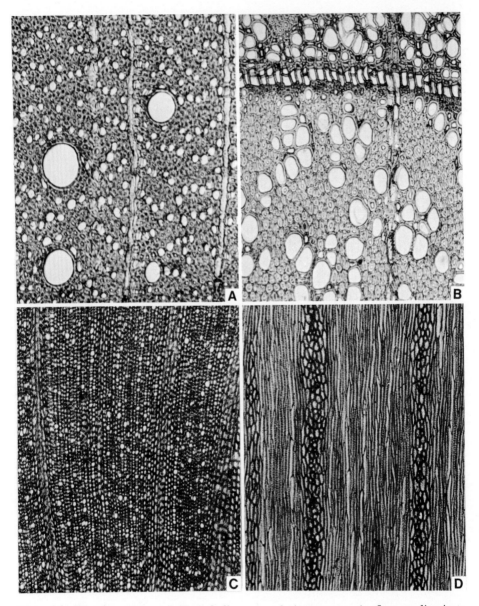

Plate 15. Wood sections of Zygophyllaceae and Asteraceae. A. *Larrea divaricata* (Zygophyllaceae); transection, vessels, fibers thick walled. B. *Artemisia arbuscula* (Asteraceae); transection, latewood with a band of interxylary cork above. C. *Loricaria thuyoides* (Asteraceae); transection, vessels and vascular tracheids are extremely narrow. D. *Loricaria thuyoides*; tangential section, pitting reveals all tracheary elements to be vessels or vascular tracheids; no libriform fibers present. All magnification according to scale shown above plate 5-A.

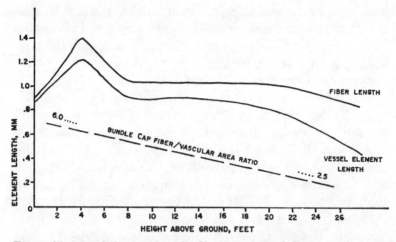

Figure 10. Vessel-element length, fiber length, and fiber/vascular area ratio according to height in a tree of *Phoenix sylvestris*. Scale at left applies to length. The line for fiber/vascular bundle area ratio has been superimposed on graph. (Based on averaged data from Swamy and Govindarajalu, 1961.)

of vessel elements adaptive. Vessels of wide diameter can be stronger if shorter, both because of the strengthening effect of the perforation-plate ring at the ends of each vessel element and for other reasons that can be deduced from structural engineering, if one is given full information about the wall characteristics of any given vessel element.

Unfortunately we have no information on variation of vessel-element length in most monocotyledons. Swamy and Govindarajalu (1961) presented data on the stem of *Phoenix sylvestris*. These data have been graphed here as figure 10. Similar patterns have been reported by Tomlinson and Zimmermann (1967) for *Sabal palmetto*. The basis for the rise and fall shown in the vessel-element length and fiber-length curves shown in figure 10 seems to me to lie in internode length. A palm stem typically begins with short internodes, succeeded by much longer internodes, followed by gradually shortened internodes at progressively higher levels. The relative abundance of mechanical elements, also entered on figure 10, shows a steady diminution from the base upward, suggesting a type of mechanical support like that of a woody dicotyledon or conifer trunk.

If internode length is the basis for the lengths in figure 10, the fiber lengths do not run contrary to the mechanical-strength properties illustrated by gymnosperm tracheids, in which there is evidence (chapter 7) that longer tracheids are mechanically stronger. From about the four-foot level in a stem of *Phoenix sylvestris*, fibers decrease in length upward to the top of the stem. This would be correlated with the fact that less bulk and weight is supported at successively higher levels. To be sure, the levels below the four-foot level would seem, if length of fibers were a criterion, to suffer diminution of mechanical strength where the requirement for strength would be the greatest. In addition, the trunk of a palm—considered in terms of *stem tissues only*—is narrowest at the base. However the base of a palm trunk is broadened only because of the numerous adventitious roots that expand the stem. Expansion by means of these mechanically strong roots is greatest at the base of a palm stem, as can be seen in the gross aspect of a palm trunk, and decreases to about the four-foot level. Thus, all patterns in a palm trunk are congruent with the greatest mechanical strength at ground level, with progressive decrease upward.

The pattern of decrease in vessel-element length with increasing height in the trunk (fig. 10) also has a correlation with the physiology of conduction. According to the tension-cohesion theory of water transport and the confirming measurements made by Scholander and his associates, higher tensions are present in upper portions of a plant. These are necessary if only to overcome the friction of tracheid and vessel-element capillarity. Shorter vessel elements, as in xeric dicotyledons, are less subject to collapse under high tensions, for reasons given in chapter 11. Therefore we would expect shorter vessel elements to be present in the uppermost portions of a palm stem, as is in fact the case.

Data on fiber length, tracheary-element length, transectional areas of tracheary tissues, transectional area of fibers, wall thickness of fibers, and data on phloem are very much needed in monocotyledons, and would provide materials for interesting correlations with the growth form and physiology of these plants. As one example, the tracheids produced by secondary activity in the woody monocotyledons seem to be notably thick walled, and frequently not accompanied by fibers, as illustrated for *Dracaena hawaiiensis* (= *Pleomele aurea* of most authors) by Tomlinson and Zimmermann (1967, 1969). Do these tracheids serve both mechanical and conductive functions by default of marked formation of extraxylary fibers? Is the amount of secondary activity in those monocotyledons with bundle-producing cambium related primarily to structure on a mechanical basis, with conductive function of the added tissue subsidiary? Or does the increase in the leafy surface of the crown bear a direct relationship to the amount of tracheids added by secondary activity?

A potentially very important issue is presented by variation in diameter of tracheary elements in monocotyledons. Cheadle (1943*a*) stated, "no evidence can be brought forward at this time to indicate that the diameter of vessels increases during specialization as measured by other characteristics." This opens the possibility that greater tracheary-element diameter, as in dicotyledons, might be related to considerations of conductive efficiency and also the existence of negative pressures in xylem. Most notably, the metaxylem of grass stems and many palm stems may compensate for the inability to form additional conductive tissue (figure 16). The 14 genera of monocotyledons that form secondary bundles would obviously be in a different category.

The transectional area of conductive tissue in a monocotyledon stem must supply any stem and leaf portions above the level of that transection. In monocotyledons with adventitious roots, the limitations of conductive tissue at any given level can be bypassed by roots that supply the stem above that level. Without adventitious roots, the conductive capacity of a monocotyledon stem is much like that of a dicotyledonous vine with little or no secondary growth (such as *Cucurbita*), which has exceptionally wide metaxylem vessels. The vessel diameter in woody dicotyledonous vines (table 13) also forms a parallel.

A rhizome type of construction is, of course, an effective way of assuring continual production of adventitious roots, and thereby by-passing the conductive limits of the finite monocotyledonous stem or the old, deactivated portions of a stem. One would expect that the widest tracheary elements would occur in vining monocotyledons and in tall forms without adventitious roots much above the ground level, or without secondary bundles. Tracheary diameter would be expected to be least in growth forms of a rhizomatous habit, in aquatics, and in short-stemmed monocotyledons. The great prevalence of the rhizomatous and acaulescent habits in monocotyledons seems clearly related to the value of by-passing older stem portions by means of adventitious roots. This limitation has undoubtedly curtailed entry of monocotyledons into arboreal habits. Only those with extremely efficient conductive systems (palms, bamboos) or xeromorphic or succulent adaptations, usually in combination with production of secondary bundles (*Aloë*, agavoids, xanthorrhoeoids) develop tallness. *Pandanus* is an exception because of its mesic habitats and optimal production and size of aerial roots. The tallest *Pandanus* species occur typically in the wettest forest. Because the number and size of roots can vary in individuals of monocotyledons at large, as can the number and size of leaves, the stems of monocotyledons can be regarded as a "bottleneck" where diameter and total transectional area of tracheary elements is decisive, unless adventitious roots intervene.

Wall thickness of tracheary elements, according to Cheadle (1943a), shows little or no correlation with trends of vessel specialization. Here again, we are dealing with a feature closely tied to habit, and each genus and species will probably be interpretable not so much in terms of vessel-element length and perforation plates as in terms of plant form, and of which tissues serve for mechanical strength. Wall thickness of vessels may be expected to be tied primarily to the characteristics of a tubular conductive system containing liquid under tension, unless tracheids (as in *Dracaena*) serve both mechanical and conductive functions.

Thus, any consideration of diversification in the habit of monocotyledons would be incomplete without consideration of xylem. Xylem structure governs the capability to exist under particular ecological conditions, and the growth forms that can exist. The compensations of leaf form, surface, stem succulence and presence of adventitious roots must be kept in mind in formulating correlations.

NYMPHAEALES

The recent discovery of vessels in *Nelumbo* (Kokasai, Moseley, and Cheadle, 1970) deserves mention because this family has been regarded as primitively vesselless, for reasons cited by those authors. Origin of vessels in *Nelumbo* is therefore to be regarded as independent of origin of vessels elsewhere in dicotyledons. The inherent interest of vessel occurrence in *Nelumbo* is not so much that weakly differentiated perforation plates do occur. In fact, there is not much difference between vessel elements of this type and tracheids in roots of other Nymphaeaceae, which have broad pit areas suitable for slow intercellular transfer of water. The feature of interest is that vessels occur in roots of *Nelumbo* (which lacks secondary growth) and this represents a monocotyledonous pattern unlike origin of vessels in woody dicotyledons. *Nelumbo* can be compared to aquatic monocotyledons such as Potamogetona-

ceae. *Nelumbo* has succulent rhizomes and grows in ponds
and ditches that can dry seasonally, so that occurrence of vessels to facilitate conduction of water to the rhizomes when
water is available is understandable. This seasonality is greater
than is characteristic of habitats of other Nymphaeales.

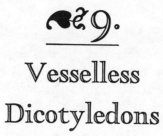

Vesselless
Dicotyledons

Vesselless dicotyledons (other than Nymphaeaceae and aquatics with reduced vascular tissue) consist of Winteraceae, Trochodendraceae, Tetracentraceae, Amborellaceae, and *Sarcandra* of the Chloranthaceae. These woody genera conform in plan to gymnosperm wood patterns, with certain notable exceptions, such as universal presence of multiseriate rays in addition to uniseriate rays (plates 4–7) and a tendency in several genera to scalariform pitting in certain areas of tracheids (plate 5-A,C; plate 6-D; plate 8-A,C,D). As in gymnosperms, pits tend to be sparse on tangential walls of tracheids (plate 4-B,D; plate 5-B,E; plate 6-B; plate 7-B). *Tetracentron* (plate 4-C,D) and *Trochodendron* (plate 4-A,B) show marked growth rings, whereas the other vesselless genera show less variation in radial diameter of tracheids (plate 5-D, plate 6-A, plate 7-A).

Average tracheid lengths in the vesselless dicotyledons show a very wide range (fig. 11) that could perhaps be extended a little by the study of additional species, but probably not by very much. The range in plant size in vesselless dicotyledons is less than in conifers, so if tracheid length is related to plant size, one would expect a more limited range of lengths in vesselless dicotyledons than in conifers.

Plant size does appear closely related to tracheid lengths in the vesselless dicotyledons. Because of the limited size of individuals in most species, wood samples from the bases of mature specimens can be obtained rather easily. I obtained a relatively small figure for average tracheid length (1,620 μ) for *Tetracentron sinense*, as did Bailey and Tupper (1918) because that sample came from a relatively small branch. A sample from the main stem might well show the longest tracheids of any vesselless dicotyledon, for it is apparently the tallest tree in the group, up to 30 meters in height (Smith, 1945). *Trochodendron aralioides* ranges from 5 to 20 meters in height (Smith, 1945).

ECOLOGY

Vesselless dicotyledons operate with the same restrictions as conifers with respect to efficiency of conductive system. Not surprisingly, many of the vesselless dicotyledons grow in company with upland tropical conifers (*Podocarpus*, *Phyllocladus*, *Dacrydium*, and *Araucaria* notably). The woody vesselless dicotyledons thus occur in highly mesic sites where efficiency of conduction is not of high selective value. They are comparable to the broad-leaved tropical conifers. As mentioned in chapter 7, a vesselless wood is compatible with a broad-leaved habit only under conditions of moist soil and minimal transpiration. We might add, to the "broad-leaved conifers" the genus *Phyllocladus*, even though its extensive photosynthetic surface is phyllocladial rather than foliar. The similarity in habitat preference by *Phyllocladus* and by *Tasmannia* (Winteraceae) in Tasmania and in New Guinea is notable; they often may be found together.

Understory habit, typical of Winteraceae, Amborellaceae, and especially *Sarcandra*, decreases transpiration rates. The understory nature of Winteraceae is obvious to those who have seen them in the field, but can also be seen in the plant heights given for the various species by Smith (1943*a*, 1943*b*). Although subalpine specimens of *Tasmannia piperita* (sensu lato) and *T. lanceolata* (Winteraceae) are relatively small shrubs with reduced leaves, none of the vesselless dicotyledons have achieved anything that could be called a microphyllous habit. Deciduousness in a moist temperate forest occurs in *Tetracentron*, and could be said to be paralleled by *Ginkgo* and, among conifers by *Taxodium, Metasequoia, Larix,* and *Pseudotsuga.* Vesselless dicotyledons, if we can judge from the assemblage alive today, have not competed successfully in the niche of canopy trees. As sub-canopy trees and shrubs, they are in the adaptive zone most suited to their probably low rates of water conduction.

Occlusion of stomata of leaves by "alveolar occluding materials" was first reported for leaves of Winteraceae by Wulff (1898), described further by Bailey and Nast (1944), and studied in detail by Bongers (1973). Bongers notes absence of alveolar material in *Bubbia perrieri* and in most entities in the genus *Tasmannia.* Bailey and Nast (1944) advanced the possibility that stomatal occlusion compensated for conductive inefficiency in Winteraceae. One might alternatively say that it has permitted the broad-leaved habit to be operative in areas subject to fluctuation in transpiration. Bongers (1973, p. 403) is skeptical of this hypothesis. However, absence of stomatal occlusion in the species of Winteraceae cited above may be a tolerable condition if plant size is small and leaf size is limited. Indeed, in Bongers' data on leaf size and habitat, one notes marked leaf-size reduction in species of *Tasmannia* from sunny habitats.

Among vesselless dicotyledons, *Trochodendron* has leaves which wilt or dry very slowly when mature (observations mine). Bailey (1953) noted that stomata in *Trochodendron* are often plugged or malformed. The deciduous nature of leaves of *Tetracentron* may provide a mechanism for the restriction of transpiration. The fact that these two genera have broad, long scalariformly pitted overlap areas on earlywood tracheids (plate 5-A,C) suggests that they have greater conductive capacity and this, in turn, may explain why they form taller trees than do Winteraceae.

Bongers (1973) notes that various vessel-bearing angiosperms have occluded stomata, as reported by Wulff (1898). However this does not, as Bongers seems to believe, vitiate the possibility that plugged stomata compensate for broad leaf form in Winteraceae and Trochodendraceae. Rather, the mechanism permits any vascular plant to reduce its transpiration rate, and thereby to widen its tolerances and extend its range. The efficacy of plugged stomata in reduction of transpiration has been shown experimentally by Jeffree, Johnson, and Jarvis (1971).

PITTING AND THE PRE-VESSEL TRACHEID

Tracheid pitting of vesselless dicotyledons does differ from that of gymnosperms. Bailey (1944a) wisely notes that the scalariform radial-wall pitting in thin-walled earlywood tracheids of *Trochodendron* (plate 5-A) and *Tetracentron* (plate 5-C) cannot be regarded apart from the circular bordered pitting of the thick-walled latewood tracheids. He also notes that stems of Cycadeoideales and very young stems of Winteraceae with prominent earlywood may exhibit this dimorphism. One cannot, Bailey implies, cite just the earlywood tracheids as hypothetical precursors of vessel elements, because "The more primitive types of vessels are diffused throughout the wood and are not in zonal arrangement" in vessel-bearing dicotyledons. Bailey (1950, p. 101) feels that, "although the plastic vesselless wood of the Winteraceae more closely approximates the type in which vessels originated, the actual ancestral forms must have contained a higher ratio of scalariform pitting than occurs in most living representatives of the Winteraceae, which exhibit evidence in the reduction of such pitting." Is this true? Bailey is probably right in regarding tracheids such as those of *Drimys winteri* (plate 7-C) as specialized where radial walls are wide enough to bear scalariform pitting, but bear instead several series of circular bordered pits (note also *Bubbia semecarpoides*, plate 8-B). Some species of Winteraceae have a few elongate pits on radial walls, but mostly circular bordered pits in which the apertures follow the helix of the pits (plate 8-D).

I have noticed, as did Bailey (1944a), that a scattering of

species of Winteraceae have scalariform pitting on the radial walls of tracheids where tracheids overlap. Among the Winteraceae I have studied, I found this condition in all species of *Bubbia* (plate 8-C) and *Zygogynum* (plate 8-A), but not in the species of *Belliolum, Drimys, Pseudowintera,* or *Tasmannia* in my wood collections. Occurrence of circular pits would maximize strength in winteraceous tracheids. Winteraceae have poorly marked growth rings or none at all (this characteristic is what Bailey means by "plastic"). Thus, they cannot form latewood compensatory in strength for weak earlywood, and all tracheids must be of about the same degree of mechanical strength. Occurrence of scalariform pitting in the overlap areas is logical, because one would expect greater conductive area on end walls of tracheids. I also found the same phenomenon in wood of *Amborella trichopoda* (plate 6-D; note the difference from the tangential walls in plate 6-C). Bailey and Swamy (1948) do report and figure both scalariform and circular pits in *Amborella* tracheids. Scalariform pitting occurs on radial walls of wider tracheids of *Sarcandra glabra* (Swamy and Bailey, 1950), from the beginning of secondary growth onwards (plate 5-F; compare this with the tangential tracheid walls in plate 5-E).

The significant feature of the above observations seems to me to be that *Trochodendron, Tetracentron, Sarcandra,* and *Amborella* have *retained the ability to form scalariform pits where tracheids wide enough to bear such pits are formed.* This capability has been retained in Winteraceae only in the overlap areas of *Bubbia* and *Zygogynum* tracheids. It is this residual ability to form scalariform pits on lateral walls and overlap areas that I find significant. We do not need to hypothesize, as Bailey does, a wood consisting mostly of scalariformly pitted tracheids to account for origin of vessels that conform to Frost's (1930a) description of a primitive vessel in a dicotyledon. A wood in which all tracheid walls bore scalariform pitting, or in which even the radial walls were all wide and bore scalariform pitting, regardless of the nature of tangential walls, would be excessively weak. As mentioned in chapter 4, wood of such a formulation existed in fossil pterido-phytes. The mechanical limitations of such wood were com-pensated for by the massive development of cortical scleren-chyma—a formula which was tenable only with microphylly, special growth forms, and unusual ecological conditions. That wood similar to that of the fossil pteridophytes was ancestral to vessel-bearing dicotyledons seems entirely untenable.

If a vessel element were, hypothetically, to be formed in the wood of *Amborella, Trochodendron, Tetracentron,* or *Sarc-andra,* it would be appreciably greater in diameter than a narrow tracheid. All of its lateral walls would be at least as wide as the radial walls of earlywood tracheids, and the walls would therefore be expected to bear scalariform pitting. Thus, we need only hypothesize a vesselless dicotyledon in which the *ability* to form scalariform pits if a tracheary element wide enough for them to occur is retained. *Sarcandra* (plate 5-D,F) demonstrates how very easily this is possible. Radial walls bear scalariform pitting throughout a year's growth. The pitting in overlap areas, in fact, is very close to approximating a perforation plate, differing merely in presence of pit mem-branes. A slight increase in diameter would be expected if a *Sarcandra* tracheid were to be transformed into a vessel ele-ment. *Chloranthus,* which is very closely related to *Sarcandra,* satisfies this description perfectly, judging from the figures of Swamy (1953) for *C. officinalis.* Thus the *Sarcandra-Chlo-ranthus* series is an example of a "non-missing link" in vessel origin. The perforation plates of *Chloranthus officinalis* fea-ture fully bordered bars (Swamy, 1953). The *Sarcandra-Chlo-ranthus* continuum is a continuum not only in taxonomic terms; the two genera have the same habit.

One need not hypothesize vessel origin only in small shrubs such as *Chloranthus.* The ability to form scalariform pits on the radial walls of wider tracheids can occur in larger shrubs and trees as well. Bailey believes that non-obligate growth rings must characterize a vesselless dicotyledon that gave rise to vessel-bearing dicotyledons. I would tend to agree with him, and concede that *Trochodendron* and *Tetracentron* do have obligate growth rings. However, there is no a priori reason why even in these woods a vessel element formed in latewood would not have scalariform lateral wall pitting, for the element would be at least as wide as an earlywood tracheid and would therefore tend to have scalariform pitting. In fact, it would be difficult to imagine a mechanism that would restrict vessel elements, once they had come into existence, to earlywood alone.

However, we can assume that a vesselless dicotyledon with non-obligate growth rings may have been ancestral to vessel-bearing dicotyledons. *Trochodendron* and *Tetracentron* and some Winteraceae have adapted to temperate regimes. However, judging from the distribution of gymnosperms with the same formula (broad leaves), the majority of vesselless dicotyledons probably existed in tropical uplands. Vessels may well have originated several times among vesselless dicotyledons. In fact, origin of vessels in Gnetales may have occurred under conditions quite different from those in dicotyledons. *Ephedra*, for example, has vessel elements with perforation plates more primitive than those of *Gnetum*. Did vessel elements originate in Gnetales in a xerophyte with the marked growth-ring phenomena of *Ephedra*? There is no reason why this is not possible.

Obviously, except for the *Sarcandra-Chloranthus* continuum, we do not have vesselless dicotyledons that have closely related vessel-bearing relatives, and both vessel-bearing and vesselless dicotyledons may be expected to have diverged, and such divergence would be expected in wood anatomy. The confinement of most pits to radial walls of tracheids of vesselless dicotyledons and conifers can be regarded as requisite for the mechanical strength of vesselless wood in a woody plant. Occurrence of a maximal amount of pitting on the radial walls of tracheids is almost certainly a requisite for conduction. The compensation for the mechanical weakening caused by pit abundance on radial walls would understandably be a paucity of pits on tangential walls. As soon as origin of vessels occurs, a division of labor provides mechanically strong tracheids, as in *Illicium* (plate 8-E, plate 9) or *Michelia* (plate 8-F). Because all walls of a vessel can bear scalariform pitting—and because tracheids in vessel-bearing woods do retain some conductive capacity—one would expect vessel-tracheid pits to feature broad areas of pitting, perhaps often scalariform, on the lateral walls of the vessels.

The above indicates that, contrary to Takhtajan (1969), paedomorphosis is definitely not needed to explain origin of scalariformly pitted tracheids, and via such tracheids, scalariformly pitted vessels with scalariform perforation plates in angiosperms. Paedomorphosis (see chapter 11) characterizes only stem succulents, rosette trees, and other herbs (Carlquist, 1962), and such plants are as antithetical to the woody habit universal in "primitive" dicotyledons as are cycads (which are a gymnospermous example of paedomorphosis). To be sure, scalariformly pitted tracheary elements in the secondary xylem of primitive woods may represent a continuation, in the sequence from primary to secondary xylem, of metaxylem scalariformly pitted elements. If only scalariform tracheids like metaxylem elements were formed, however, a wood too weak for arborescence or even shrubbiness would be formed.

Dicotyledons with Primitive Vessels; Gnetales

Viewed from the standpoint of radiation into and exploitation of habitats by dicotyledons, origin of vessels is perhaps the most significant initial factor. To appreciate this, we have only to compare the geographical distribution of vessel-bearing dicotyledons (or just those with primitive vessels) to the distribution of the vesselless dicotyledons. Vesselless dicotyledons may well have been more widespread than they are today; but if so, they have yielded rather quickly to the rather marked superiority of vessel-bearing dicotyledons. Differentiation in secondary xylem between a perforate conducting element and an imperforate mechanical element (both of which can be patterned according to characteristics suited to a given habit or habitat) is the prime functional explanation for success of this morphological plan. Additionally, the fact that there is an ontogenetic dimension in woody stems means that the woody vessel-bearing dicotyledons have all the flexibility of monocotyledons without the limitations imposed by a lack of cambial activity. Vessel-bearing dicotyledons have a greater capability for diversity in habit than gymnosperms, without the gymnospermous disadvantage of a single type of tracheary element.

In strictly morphological terms, the early stages in the evolution of the vessel element have been describd by Frost (1930a). The question that must now be answered is why primitive vessels persist at all, and just which characteristics of primitive vessels and tracheids have selective values. A second question, which is a corollary to the first, is under what ecological conditions woods with primitive vessels form a successful xylary plan.

QUANTITATIVE DATA ON PRIMITIVE DICOTYLEDONS

For the purpose of analyzing the initial stages in origin of vessel elements, data were gathered from the wood of 28 dicotyledons. These species were selected because each has more than 10 bars per perforation plate, diffuse axial parenchyma, and tracheids (rather than fiber-tracheids or libriform fibers) as the imperforate elements of the secondary xylem. The species are listed in the legend for figure 12. They represent 20 families, and are not a tightly knit group according to any of the various classification systems.

Data on these 28 species are represented in figures 12, 13, and 14-A, and form one of the paired bars in figure 11. We can immediately see that vessel diameter is not correlated with vessel-element length (fig. 14-A). The spread of points would have been even more scattered if species with more specialized woods had been included. In any case, nothing approaching a straight line is discernible, although there are ultimate limitations. Vessel elements that average greater diameter than length are probably impossible (at least with scalariform perforation plates), as are excessively long vessel elements, which would be physically weak and subject to collapse unless very thick walled.

More significantly, figure 14-A gives us one of several reasons (others are given later) for believing that vessel diameter

Figure 11. Tracheary-element length for groups of woody vascular plants, based on averages for particular species. Bars indicate range of averages, broad vertical line indicates average of average lengths for species in each category. The 28 dicotyledons on which the upper pair of bars for vessel-bearing dicotyledons is based are listed in the legend of figure 12.

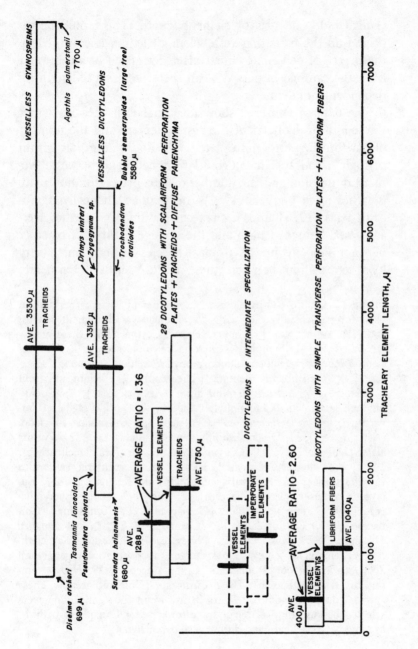

evolves independently of element length. This would be expected on the basis that vessel-element length is governed by the degree of cell enlargement after derivation of an element from the cambial initial. We can therefore analyze length independently of diameter.

The data of figure 12 show no correlation between either the length or diameter of the vessel element and the number of bars on the perforation plate. One should note that within a single wood specimen, correlation between the vessel-element dimension and the number of bars per perforation plate is rather loose (Tabata, 1964) and is not statistically significant. Baas (1973) finds a close correlation between the average vessel-element length and the average number of bars per perforation plate in the tropical species of *Ilex*, but a very loose correlation between these features in the temperate

Figure 12. Graphs comparing average number of bars per perforation plate for particular species with (A) average vessel element length; and (B), average vessel diameter. The species on which these graphs are based are: 1, *Symplocos tinctoria* (Symplocaceae); 2, *Sphenostemon lobospora* (Sphenostemonaceae); 3, *Schima noronhae* (Theaceae); 4, *Ternstroemia dentata* (Theaceae); 5, *Illicium anisatum* (Illiciaceae); 6, *Aextoxicon punctatum* (Aextoxicaceae); 7, *Cercidiphyllum japonicum* (Cercidiphylphyllaceae); 8, *Curtisia faginea* (Cornaceae); 9, *Cyrilla racemiflora* (Cyrillaceae); 10, *Saurauia subglabra* (Actinidiaceae); 11, *Dillenia philippinensis* (Dilleniaceae); 12, *Wormia alata* (Dilleniaceae); 13, *Distylium racemosum* (Hamamelidaceae); 14, *Sycopis dunnii* (Hamamelidaceae); 15, *Carpenteria californica* (Saxifragaceae); 16, *Pennantia cunninghamii* (Icacinaceae); 17, *Eucryphia cordifolia* (Eucryphiaceae); 18, *Eucryphia lucida* (Eucryphiaceae); 19, *Cornus alternifolius* (Cornaceae); 20, *Cornus mas* (Cornaceae); 21, *Cornus nuttallii* (Cornaceae); 22, *Nyssa aquatica* (Nyssaceae); 23, *Nyssa sylvatica* (Nyssaceae); 24, *Pentaphylax clethroides* (Pentaphylacaceae); 25, *Strasburgeria robusta* (Strasburgeriaceae); 26, *Hedyosmum bonplandianum* (Chloranthaceae); 27, *Hedyosmum nutans* (Chloranthaceae); 28, *Ascarina* sp. (Chloranthaceae). These species (documentation on labels in my wood slide collection) were selected because they all have: scalariform perforation plates with 10 or more bars; tracheids; and diffuse axial parenchyma.

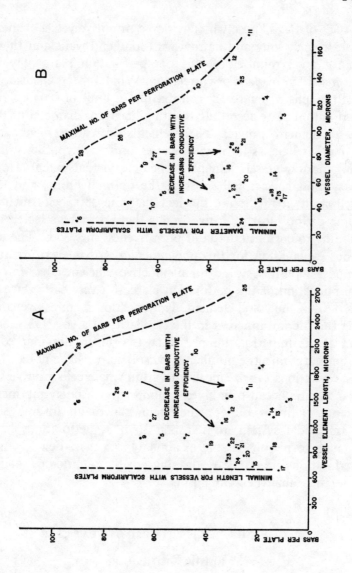

species of *Ilex*. Those who envision a primitive vessel element as extremely long and narrow with long end walls and therefore a great number of bars per perforation plate may be surprised by this lack of correlation. What appears instead, in both graphs of figure 12, is a maximum limit of bars on the perforation plate according to either length or diameter of the vessel element. Longer vessel elements do tend to have more numerous bars, as do narrower ones, and the hypothesis of a long, narrow vessel element is operable to that extent. However, most species fall short of the optimal limits, and fall well short of it, despite the fact that the 28 species include some of the most primitive woods that could be selected according to current criteria of primitiveness in woods. The obvious explanation for this phenomenon is that even in these woods, a positive value for conductive efficiency has forced simplification of the perforation plate at a very early stage in phylesis of the vessel element. Translocation requires overcoming the friction inherent in the capillarity of a vessel element. A scalariform perforation plate increases this friction markedly, in proportion to the number of bars. Also, we have noted that air cavitations can be localized within single cells in a vessel-less xylem, because bubbles cannot pass through pit membranes. If this is, under some circumstances, an advantage of a tracheidal system, it is lost as soon as perforations large enough to permit passage of air bubbles occur. Fewer, larger perforations are no more advantageous than numerous smaller ones in localizing air embolisms.

ECOLOGY OF WOODS WITH PRIMITIVE VESSELS

Floristic Distribution

The reader who knows the ecology of the species listed in the legend for figure 12 will immediately notice that most of the species occur in as highly mesic situations as could be imagined, with a minimum of seasonal water stress. Only in a few (e.g., *Carpenteria californica*) could one cite summer heat and low humidity as probable factors for more rapid transpiration and higher conductive rates.

One may hypothesize that rapid departure from the tracheid-like configuration of vessels is due to the probability that under almost any circumstances in which dicotyledons could live, greater conductive efficiency is of selective value. The rapidity of departure—or alternatively, the preadaptational advantage of the simple perforation plate—seems obvious when we consider what a small proportion of dicotyledons at large have scalariform perforation plates.

Southern California.—A mediterranean type of climate dramatizes the selective value of a simple perforation plate where strong seasonal fluctuation in water availability occurs. In the native flora of southern California, I know of only five genera in which scalariform perforation plates occur: *Cornus*, *Alnus*, *Umbellularia*, *Garrya*, and *Ribes*. Of these, *Cornus* (*C. nuttallii*) has the most primitive vessel elements (Adams, 1949), and is narrowly restricted to the most mesic sites available: deep shady montane canyons. The other four genera—which have few bars per plate, or simple plates mixed with scalariform—occur in mesic sites. *Umbellularia* occurs in canyons close to streams, as does *Alnus*. *Ribes* is typically an understory element, and most frequently occurs on slopes of shady canyons, sometimes in streambeds. *Garrya* is a chaparral element, but its presence in rocky montane streambeds in southern California demonstrates that it occurs in relation to underground water. The vast bulk of the woody elements in the southern California flora—*Quercus*, *Salvia*, *Arctostaphylos*, *Rhus*, *Adenostoma*, *Artemisia*, *Encelia*, etc.—has simple perforation plates exclusively. Thus, with the few exceptions noted, southern California has a "simple perforation plate flora" and one can hypothesize that only by virtue of this type of wood are the native species capable of adaptation to this climate.

The "old wet forests."—On the contrary, some tropical and temperate areas have a climate of year-long moisture availability. Those areas where such climates have prevailed for long periods of time may be expected to have species with more primitive mesomorphic species. Such areas can be characterized as tropical upland and wet temperate forest, and include such areas as montane New Caledonia, Chiapas (Mexico), the Malesian highlands, the southeastern United States, Japan, and summer-wet portions of China and Europe.

Versteegh (1968) contrasted the montane Javan flora with the lowland Javan flora, noting that in selected genera represented in both regions, species from the uplands have more numerous bars per perforation plate than their lowland counterparts. Baas (1973) presented percentages, utilizing in part data from Kanehira (1921a, 1921b, 1924), Janssonius (1906–1936) and Greguss (1959). These percentages have been presented here as table 8. Table 8 shows clearly that both tem-

TABLE 8

Percentages of Genera of Dicotyledons in which Woods Have Scalariform Perforation Plates, by Floras

Area	Percent
Japan	32.6
Temperate and subtropical Europe	23.0
Taiwan	19.4
Tropical Java	13.0
Philippines	5.7

perate and tropical forests qualify as refugia for species with primitive wood structure, but only if summer moisture and cloud cover are present. To the extent that lowland species are included in the calculations of table 8, percentages decrease.

Hawaiian Islands.—The Hawaiian Islands present an exceptional situation. The only woody genera I can cite in this flora that have scalariform perforation plates are *Ilex, Perrottetia, Cheirodendron*, and *Eurya* (Brown, 1922). This would be approximately three percent of the genera of dicotyledons in the Hawaiian flora—a very low percentage for an area with a high average rainfall. The tendency of the Hawaiian flora to be a "simple perforation plate flora" can be seen in terms of individual groups. Genera of Hawaiian Araliaceae other than *Cheirodendron* have simple perforation plates. Genera and families in which Hawaiian species have simple perforation plates, but non-Hawaiian species have scalariform perforation plates include *Styphelia* (Epacridaceae) and *Cryptocarya* (Lauraceae). If one consults the distribution maps for Pacific genera such as those of Steenis (1963) and Steenis and Balgooy (1966), one finds many families typical of Malesian forests characterized by scalariform perforation plates do not reach Hawaii. Many do not reach other high Pacific volcanic islands. These include Alangiaceae, Celastraceae (most), Cornaceae, Dilleniaceae, Eucryphiaceae, Hamamelidaceae, Illiciaceae, Nyssaceae, Rhizophoraceae, Saxifragaceae (some subfamilies), and Styracaceae. The chief forest tree of the Hawaiian wet forest, *Metrosideros*, has simple perforation plates. Correlatively, no conifers are native to the Hawaiian Islands, although some (e.g., *Araucaria*) have naturalized there. The lack of conifers suggests one of the explanations for the paucity of scalariform perforation plates in the Hawaiian flora: the "scalariform perforation plate flora" (trees of upland continental tropics) has poor dispersibility. The plants with the best dispersibility, worldwide, are typical of drier and unstable habitats. Asteraceae are a good example. In the Hawaiian flora, Asteraceae have adapted secondarily to wet forest. This give us the second reason for the low percentage of scalariform plates: the phylads that have evolved into wet Hawaiian forest are not disadvantaged by simple perforation plates in wetter situations, and the scalariform condition, of course, cannot be reconstituted in phylads where it is absent.

Lowland Tropics.—The above explanation appears adequate for the Hawaiian flora. Does it apply to floras of tropical lowlands, which have a notably low percentage of genera with scalariform perforation plates? Baas (1973) states, "The attempt to bring these facts into harmony with the hypothesis that the tropical lowland favors primitive wood structure would involve fantastic speculation, like the assumption of an evolution outside the hot tropical environment for those families now constituting the bulk of the tropical flora." We do not need to assume migration of groups with specialized xylem into the hot tropical lowlands. Rather, in contrast with an implication of the above quotation, we may assume that the hot tropical lowlands do not favor woods with scalariform perforation plates, and that woods with simple perforation plates have evolved autochthonously within tropical lowlands. To claim otherwise would be to claim that species with scalariform perforation plates grow under conditions to which they are not adapted. Hot tropical lowlands are subject to wide fluctuations of soil moisture (because of monsoon rainfall patterns, for example). Periods of high insolation would result in rapid transpiration and high conductive rates. With the exception of mesic pockets, tropical lowlands could be characterized as "simple perforation plate floras." This accords with Versteegh's (1968) data. The problem Baas (1973) sees in this connection is evidently based on the fact that *Ilex* species in tropical lowlands have many bars per perforation plate. Baas (p. 220) hypothesizes that the tropical lowland species of *Ilex* may represent reversion (that is, a secondary increase in the number of bars per perforation plate). In an addendum (p. 251), Baas considers that lowland tropical *Ilex* runs counter to the general trend and may be primitive. I would say that primitive xylem characteristics probably have been retained in *Ilex* species in tropical lowlands (which may have more specialized floral characteristics, as Baas implies). However, these xylem characteristics are perhaps retained because within the tropical lowlands, *Ilex* may occupy more mesic sites. *Ilex* seems often an understory element in the tropical lowlands. Also worth noting is that there are relatively few species of *Ilex* in tropical lowlands; the genus is much better represented in moist tropical uplands where cloud cover and rainfall provide more markedly mesic habitats.

The above problems show that the ecological preferences of species must be known if meaningful correlations are to be made between wood anatomy and factors influencing the evolution of wood. Microclimates are often involved, and broad generalizations are not meaningful in this regard. Unfortunately, those who collect woods are often not the same individuals as those who study wood anatomy; botanists who are acutely familiar with the ecological preferences of species are often field botanists who do not study wood anatomy.

South Africa.—The flora of South Africa is predominantly a "simple perforation plate flora" because of its mediterranean climate, especially in the Cape Province. Understanding of microclimatic pockets is important, because woods with scalariform perforation plates occur in species restricted to areas that represent but a fraction of the land surface. In the southwestern Cape (the area formed by the Table Mountain Sandstone) trees characterize riparian habitats. Some of these trees, such as *Curtisia* and *Cunonia*, have scalariform perforation plates. The characteristic vegetation of the Cape sandstone mountains consists of shrubs (*Erica, Penaea,* Fabaceae, Proteaceae) to a large extent. Some shrubs, however, occupy special sites; most notable among these shrubs is a series of families perhaps closely related to each other: Bruniaceae, Geissolomataceae, Grubbiaceae, and Roridulaceae. These four families occur only in the most mesic places of montane Cape Province: seeps, marshes, streamsides, cool south-facing slopes, rocky outcrops, and rocky summits often cloaked by clouds. The four families seem old elements in the South African flora, and have probably survived only in limited mesic pockets. All four families have scalariform perforation plates in secondary xylem. Reduction in leaf surface, perhaps in proportion to windiness as well as sun during summer months, characterize Bruniaceae and Grubbiaceae. Leaves of Roridulaceae are covered with a sticky coating that slows desiccation, whereas leaves of *Geissoloma* are hairy when young, but bear a thick cuticle upon maturity.

As in the Hawaiian flora, species with simple perforation plates also occupy mesic sites in the South African flora. The conductive characteristics of such vessels in no way hinder adaptation to moister habitats. I believe that old mesic groups that have had an unbroken history of occupation of mesic sites can be distinguished from secondary invaders with sufficient study. There can be little doubt that *Ilex*, which occurs both in the Hawaiian Islands and in Cape Province, belongs to a family that is primitively mesic and has remained so. On the other hand, some of the woody species of Asteraceae are restricted to mesic habitats both in the Hawaiian Islands and Cape Province, and these are secondary mesophytes, for Asteraceae at large are not mesophytes and all Asteraceae have simple perforation plates.

Analyses in detail of various floras would be interesting, because wood anatomy could be an important tool to discern the history of taxonomic groups, the floristic history of regions, and interreactions between the two.

Systematic Distribution

Instead of looking at a flora, one can look at woody dicotyledons as a whole to determine the ecological conditions under which woods with scalariform perforation plates tend to exist. I have, therefore, compiled a list in terms of habit and habitat. The listing of families with scalariform perforation plates by Metcalfe and Chalk (1950, p. 1349) proved helpful.

In compiling the list below, I noted that only a small number of families have long scalariform perforation plates (usually more than 20 bars) exclusively. These families are indicated in table 9 by "LS". Most of these families consist of one or several genera. The only exception to this is Theaceae, which has about 25 genera. The fact that so many of the families with long scalariform perforation plates are monogeneric is suggestive either of a lack of radiation in these groups or of their relictual nature, or both. Obviously they have not undergone radiation recently, and this seems important because groups with simple perforation plates have probably undergone more radiation in recent times. Radiation of the phylads with primitive wood has been forestalled because groups with specialized wood have occupied the mesic sites to which the species with primitive xylem are restricted.

In addition to families with characteristically more than 20 bars per perforation plate, I have included in the list families with a range from many bars in vessel elements of some genera and species to only a few or no bars in others. These families are indicated by "R". These families have more numerous genera than those with long scalariform plates. This indicates that radiation into a wider range of sites is accompanied by, if not permitted by, simplification in the perforation plate.

Families that characteristically have scalariform perforation plates, but with fewer than 20 bars, are also included in table 9. These families are designated by "F". The total number of families that qualify for listing in table 9 is still remarkably small when compared to the families (some with large numbers of genera) that have simple perforation plates, or simple plates plus plates with one to a few bars. Families indicated "in part" have some genera that are trees, others that are shrubs, and thus are listed more than once.

Obviously these listings represent some oversimplification, and a case could be made for additions and subtractions from the list. Empetraceae does not fall easily into the two categories of shrubs given, because it grows on marine bluffs with a foggy and windy, but humid regime. Some of the "Trees of moist forest" are in tropical upland cloud forests, some are in temperate forests with wet winters and humid summers (southeastern U.S., Japan), and some are in lowland hot tropics (Amazon Basin). The same is true of shrubs.

Illicium: a primitive mesophyte.—Illicium cambodianum (plate 9) and other Illiciaceae are typical of woody plants with long scalariform perforation plates. *Illicium cambodianum* occurs as a shrub or small tree in Malesian montane forests. Although rainfall shows seasonality in this region, humidity is high at almost all times. Montane regions receive cloud cover and showers even during the dry season. Soil moisture is rather constant. For a plant under this regime, the point of greatest vulnerability is during the emergence of new growth.

TABLE 9

Dicotyledon Families with Scalariform Perforation Plates, Arranged According to Habit and Habitat

Trees of moist forest, some with tendencies toward understory:	Eucryphiaceae (R)
Actinidiaceae: Saurauia only (LS)	Eupteleaceae (LS)
	Fagaceae (R,F)
Aextoxicaceae (LS)	Flacourtiaceae (in part) (R)
Alangiaceae (LS)	Goupiaceae (LS)
Aquifoliaceae (LS)	Hamamelidaceae (F)
Betulaceae (LS)	Humiriaceae (LS)
Canellaceae (R)	Icacinaceae (in part) (R)
Celastraceae (in part) (R)	Lacistemaceae (LS)
Cercidiphyllaceae (LS)	Lauraceae (F,R)
Chloranthaceae (in part) (LS)	Lecythidaceae (R)
Clethraceae (LS)	Magnoliaceae (F)
Cunoniaceae (F)	Monimiaceae (R)
Cyrillaceae (LS)	Myristicaceae (R)
Daphniphyllaceae (LS)	Nyssaceae (LS)
Dilleniaceae (except scandent species) (LS)	Octoknemataceae (F)
	Olacaceae (R)
	Pentaphylacaceae (LS)

Saxifragaceae (including Hydrangeaceae, etc.) (in part) (R)
Scytopetalaceae (F)
Staphylaeaceae (in part) (LS)
Styracaceae (in part) (F)
Symplocaceae (LS)
Theaceae (LS)
Violaceae (in part) (R)

Mangroves:
Rhizophoraceae (F)

Shrubs of moist forest:
Aquifoliaceae (in part) (LS)
Buxaceae (LS)
Caprifoliaceae (in part) (R)
Celastraceae (in part) (R)
Chloranthaceae (in part) (LS)
Cornaceae (in part) (R)
Epacridaceae (R,F)

Ericaceae (in part) (R)
Eupomatiaceae (LS)
Flacourtiaceae (in part) (R)
Icacinaceae (in part) (R)
Illiciaceae (LS)
Myricaceae (F)
Paeoniaceae (F)
Saxifragaceae (in part) (R)
Stachyuraceae (R)
Staphylaeaceae (in part) (LS)

Shrubs of non-forest habitats, but with underground water and/or shady exposure:
Bruniaceae (LS)
Ericaceae (in part) (R)
Geissolomataceae (LS)
Grubbiaceae (LS)
Myrothamnaceae (LS)
Roridulaceae (LS)
Saxifragaceae (in part) (R)

If, at this time, an unusually sunny day occurs, transpiration can be rapid and foliage can wilt (plate 9-A). Few plants typically wilt in their native habitats. The wilting of *Illicium cambodianum* seems very likely related to its xylem configuration. Scalariform perforation plates have bordered bars and narrow perforations that offer impedance to rapid flow (plate 9-B,C). At the time the photograph shown in plate 9-A was taken, soil was wet. Other species in this locality were not wilting, but all individuals of *I. cambodianum* were wilting, and none showed any indication of pathology. By late afternoon, these individuals had recovered normal turgidity. Older leaves on these plants were not wilted.

Families with scalariform and simple perforation plates.—In those families of table 9 designated "R", the species with simple perforation plates or few bars per plate can be expected to occupy sites with less constant supply of moisture, and with marked rise and fall of humidity. Examples can be indicative. In Violaceae, the genera with exclusively scalariform perforation plates are *Alsodeia, Amphirrhox, Gloeospermum, Leonia, Papayrola,* and *Rinorea.* These are trees or shrubs of tropical forest, chiefly in the Amazonian region and other portions of tropical America. Among the genera of Violaceae with simple perforation plates are *Agatea* (a climbing shrub from New Caledonia and Fiji), *Anchietia* (a climbing shrub, South America), *Hybanthus* (herbs to shrubs of exposed locations, tropics and subtropics of the Old World and New World, including temperate Australia and the Andes), *Hymenanthera* (shrubs of exposed localities, New Zealand and Norfolk Island), *Melicytus* (shrubs, often in exposed locations, New South Wales, New Zealand, Lord Howe I., Norfolk I.), and *Viola* (chiefly herbaceous except in Hawaii and Chile; herbs of seasonal temperature areas or of tropical and subtropical mountains).

In other families such as Ericaceae, these distinctions are equally marked. *Arctostaphylos,* a genus well developed in the mediterranean climate of California, is characterized there by simple perforation plates. Likewise, the Ericaceae of summer-dry areas of Cape Province, South Africa, have simple perforation plates. Ericaceae of moist forest, such as *Rhododendron, Kalmia, Pernettya, Menziesia,* etc., have exclusively scalariform perforation plates.

Vines and lianas.—Scandent plants have not been listed in table 9. The number of lianoid plants with scalariform perforation plates is very small, but requires discussion. The vining mode of construction is associated with wide, short vessel elements. This tendency accelerates loss of bars on the perforation plate. For example, the family Schisandraceae is a lianoid family closely related to Illiciaceae. In comparison to Illiciaceae (plate 9-C,D), *Schisandra* shows specialized vessels with fewer bars (15 or fewer); a second genus of Schisandraceae, *Kadsura*, has seven or fewer bars per plate.

Woody Dilleniaceae have long scalariform perforation plates. However, three genera of Dilleniaceae that are scandent shrubs have simple perforation plates: *Davilla*, *Doliocarpus*, and *Tetracera* (Metcalfe and Chalk, 1950).

Simplification of the perforation plate can be seen within a single stem. *Actinidia*, a scandent genus, has vessel dimorphism: wider vessels with simple perforation plates, and narrower vessels with scalariform perforation plates (Metcalfe and Chalk, 1950). The only scandent genus with consistently numerous bars on perforation plates is *Austrobaileya* (Bailey and Swamy, 1949). In *Austrobaileya*, as shown in plate 11-A, B,C,D, wider vessels have narrower bars, and wider perforations (as few as 10 bars); narrower vessels have more numerous bars, separated by narrow perforations, and bars are prominently bordered. Thus, in *Actinidia* and *Austrobaileya*, we see a simplification of the perforation plate in proportion to the conductive efficiency of a vessel element. Retention of numerous bars on a wide vessel element of a liana would destroy the decrease in friction achieved by a wider diameter; presence of a few, narrow bars does not decrease friction greatly, and may contribute to increased mechanical strength of the vessel. However, bars on perforation plates of lianas seem retained only under the most mesic conditions, where conductive rates may be expected to be slower.

Mangroves.—Mangroves provide a special case. Although Rhizophoraceae are the only group of mangroves with scalariform perforation plates, the plates have relatively few, wide, and bordered bars (plate 14-D). One might not expect mangroves to represent an ecological type where scalariform perforation plates would persist. Scholander, Hammel, et al. (1965) have shown very high negative pressures in the mangroves. These high negative pressures are necessary to counter the osmotic pressure of seawater. However, these pressures do not necessarily signify high conductive rates. One might expect rather steady conductive rates in mangroves because of the unlimited supply of water at their roots. This would provide a rather mesic situation, but periods of strong illumination undoubtedly induce a rise in transpiration and conductive rates. However, the strong negative pressures in mangrove stems require vessels resistant to collapse. Vessels in mangrove woods are rather thick walled, as shown by *Ceriops* (plate 14-C). The few but thick bars on perforation plates of *Ceriops* (plate 14-D) would seem ideal to resist collapse in vessels under tension.

Herbs.—The vast majority of herbaceous dicotyledons have simple perforation plates in secondary xylem. The strong seasonality of temperate herbs seems inevitably related to this pattern. Only a few species can be said to have retained scalariform perforation plates from non-herbaceous ancestry. *Paeonia* can be cited, but only a few bars per plate characterize this genus. A better example is *Cornus canadensis*, one of the few herbaceous members of its genus. *Cornus canadensis* has scalariform perforation plates, but with fewer bars per plate than the woody species of *Cornus* (Adams, 1949). Another probable example is *Heliamphora*, a sarraceniaceous genus that forms a small amount of secondary xylem. In secondary xylem, perforation plates are long and scalariform. In the three above examples, retention of the primitive perforation plate despite the herbaceous habit is understandable because these belong to genera, families, or orders in which primitive xylem is characteristic (Sarraceniaceae is probably Thealean). More importantly, in all three examples, mesomorphy is retained. One would hypothesize that in each of these examples, there has been an unbroken history of mesomorphy.

Pentaphragma (plate 10) presents a different problem. This genus, placed in a family by itself, Pentaphragmataceae (Airy-Shaw, 1941) is generally conceded not to be primitive in most respects, but to be specialized in many respects, and perhaps nearest Campanulaceae. However, secondary xylem of *Pentaphragma* has exclusively scalariform perforation plates with numerous bars on vessel elements (plate 10-D,E). Even in the wood itself, this seems contradictory to the rayless condition (plate 10-C), since raylessness has been considered a specialized characteristic (Barghoorn, 1941). However, ontogenetic considerations in the secondary xylem of *Pentaphragma* probably explain this seeming paradox. One must remember that in contrast to secondary xylem, primary xylem of a number of herbaceous dicotyledons consists of vessels with scalariform perforation plates, or a mixture of such vessels and tracheids (Bierhorst and Zamora, 1965). A number of specialized dicotyledons have only vessels with simple perforation plates in primary xylem (Bierhorst and Zamora, 1965, p. 701). However, Campanulaceae do have scalariform perforation plates in primary xylem. For example, the protoxylem-to-metaxylem sequence given for *Lobelia cardinalis* by Bierhorst and Zamora is tracheids, followed by tracheids plus vessels with scalariform perforation plates. If paedomorphosis (Carlquist, 1962) extends such a pattern into secondary xylem, vessels with scalariform perforation plates would result. This is precisely what must have happened in *Pentaphragma*. *Pentaphragma* is rayless (plate 10-C); raylessless is also an expression of paedomorphosis (Carlquist, 1962). The peculiar growth form of *Pentaphragma* (plate 10-A) accords with paedomorphosis. It is an herb with a prostrate stem, giving rise to successive erect shoots of limited (*ca.* 1 ft.) height. It has a succulent stem, and leaves are also succulent. Stem succulence is a condition under which paedomorphosis occurs (Carlquist, 1962). Equally significant, *Pentaphragma* is an understory element in the wettest places of Malesian rain forests. Under these highly mesic conditions, scalariform perforation plates are not disadvantageous. The closeness of *Pentaphragma* to Campanulaceae is suggested by presence of stem endodermis, a feature of a few highly specialized dicotyledon families (e.g., Asteraceae). In those families, libriform fibers are present. In *Penta-*

phragma, septate tracheids are the imperforate elements. Perhaps *Pentaphragma* xylem is somewhat more primitive than that of its relatives both in its tracheids and in its primary xylem vessels.

THE PRIMITIVE VESSEL SYNDROME

Under what circumstances can a wood with scalariform perforation plates be a successful plan? A series of interrelated factors is involved in these woods. Any given wood with scalariform perforation plates will not represent all of the below features, because some "solutions" to the problem of conductivity versus mechanical strength are mutually exclusive.

1. The pressure for removal of bars on the perforation plate is minimal in environments where slow conductive rates suffice for successful plant growth. Although a variety of ecological conditions can satisfy this description, a combination of constantly moist soil and high humidity is the most frequent such environment, as in habitats of Cornaceae and Nyssaceae.

2. Long vessel elements, and therefore long fusiform cambial initials, are of positive value for conductivity because fewer plates along the length of a vessel (a series of vessel elements) would provide less impedence. Shortening of fusiform cambial initials, and therefore vessel elements, without concomitant simplification of the perforation plate would increase impedance of a wood. That primitive vessel elements are longer and also narrower that those of dicotyledons at large is shown by the curves of figure 13. To be sure, narrowness of vessels provides greater friction to conduction (Rivett, 1920), but vessels are wider than tracheids, and should be viewed in that regard. Also, as indicated above for vines, an increase in vessel diameter without concomitant simplification of the perforation plate would negate the value of increased diameter.

Long fusiform cambial initials, when derivatives mature into vessel elements, dictate long overlap areas for vessel elements. Long overlap areas have numerous perforations which, collectively, form a perforation area as great or greater than

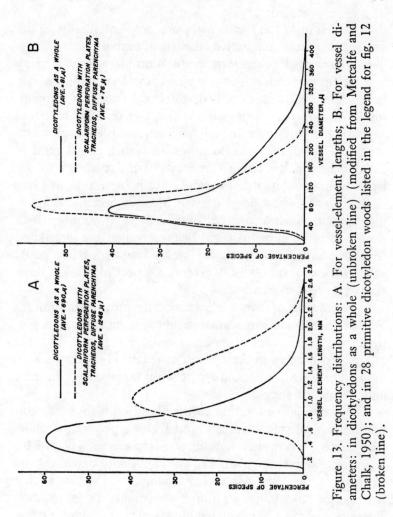

Figure 13. Frequency distributions: A. For vessel-element lengths; B. For vessel diameters: in dicotyledons as a whole (unbroken line) (modified from Metcalfe and Chalk, 1950); and in 28 primitive dicotyledon woods listed in the legend for fig. 12 (broken line).

that of a simple transverse plate are not, potentially, disadvantageous where conductive rates are not rapid.

3. Vessels with scalariform perforation plates are, in many species, thin walled and low in mechanical strength. As suggested later, short, thick-walled vessels are strong in comparison. However, bars on a perforation plate do offer potential support and resistance to collapse where negative pressures exist in a vessel, or where torsion of a branch might tend to collapse vessels. Even thin bars may provide resistance to collapse. A long oblique perforation plate is inevitable, as mentioned above, where long overlap areas between vessel elements occur. Such a long perforation plate, if occupied by a simple perforation plate, might be more vulnerable to collapse than a scalariform perforation plate. A long perforation plate, as in *Illicium* (plate 9-B,C) extends support over a long portion of a vessel element.

4. Narrowing of bars phylogenetically provides minimization of conductive impedance, while still retaining the possibility of mechanical support mentioned above. Notably attenuated bars, as in *Austrobaileya* (plate 11-C) and *Cercidiphyllum* seem a compromise between conductive efficiency and mechanical support.

5. An alternative solution to the above problem is the formation of relatively few but wide bars on a perforation plate. This can be seen in such families as Magnoliaceae (plate 8-F) and Rhizophoraceae (plate 14-D). This would maximize the strength of the bars, while also maximizing the size of the perforations and thereby minimizing conductive impedance. Presence of only a few bars could still play an important role in preventing the collapse of a vessel under negative pressures. This may help to explain why there are an appreciable number of woods, such as many Lauraceae, in which "vestigial" scalariform plates occur.

6. The presence of borders on the bars may enhance their mechanical strength, and therefore may be of positive value as long as the perforations are not narrowed to such an extent that

impedance is markedly increased. Borders on perforations characterize many woods with scalariform perforation plates, such as *Illicium* (plate 9-C,D), *Pentaphragma* (plate 10-E) and *Ceriops* (plate 14-D).

7. Some scalariform perforation plates have attenuated bars without borders in the central portions of the bars, but borders at the base of the bars (the tips of each perforation). This can be seen, for example, in *Austrobaileya scandens* (plate 11-D). Borders at the bases of bars potentially offer a form of mechanical strength. This can be compared, in structural engineering, with diagonal supports at the base of a member, forming a Y-shaped structure, that give greater rigidity. Retention of borders at the base of bars thus minimizes loss of mechanical strength of the bars.

8. Baas (1973) demonstrates a decrease in the number of bars on the scalariform perforation plate with increasing latitude in *Ilex*. This trend very likely could be demonstrated in other genera and families. I suspect Baas would agree that the climatic conditions underlying the latitudinal correlations are the important factor. I would point to a higher degree of fluctuation in transpiration and therefore conductive rates in higher latitudes. Conductivity is more rapid during the spring months, and thus provides a selective value for loss of bars on perforation plates. This can be seen in Cornaceae (Adams, 1949), in which the far northern species *Cornus canadensis* has relatively few bars. The most important factor in retention of the scalariform plate in woods of the temperate zone appears to be the occurrence of humid summers with occasional rains. The tropical uplands can be considered like the humid summer of the temperate zone extended to a year-long regime, and thus the high proportion of scalariform perforation plates in tropical uplands is understandable. Dry soil and low humidity during the summer months in the temperate zone seem to make the scalariform perforation plate of negative selective value.

9. Bands of wall material interconnecting bars can increase

mechanical strength, but at the loss of conductive ability. An example is shown here for *Eupomatia laurina* (plate 11-E). Similar conformations can be seen in Canellaceae (Wilson, 1960) and in *Carpenteria* of the Saxifragaceae (Carlquist, 1961*a*), where bars are forked, or where there are several perforations across the width of a vessel where a single perforation would be expected. All of these types of increase in bar interconnection appear to be modifications of the ordinary scalariform perforation plate, as hypothesized earlier (Carlquist, 1961*a*, p. 41). So-called multiperforate plates in dicotyledons were also cited there as such a possible modification. Interconnections between bars were mentioned in scalariform pitting of tracheids in *Sigillaria* in chapter 4. All of these phenomena appear to suggest situations where retention of the mechanical strength of the perforation plate is of greater value than the selective force for rapid conductivity. The latter may be maintained by a greater number of vessels per unit transection.

10. Thin-walled vessels with an angular transection, as in *Illicium* (plate 8-E) are probably of low mechanical strength compared to vessels circular in transection. If an angular vessel has scalariform lateral wall pitting, however, wall thickening is uninterrupted, and this would provide a maximal-strength solution for the scalariform pitting configuration. The angular configuration is probably based on the narrowness of vessels. Where vessels are not much wider than tracheids, the angular tracheid-like configuration is often attained, even in specialized dicotyledons, such as certain lobelioids (Carlquist, 1962, 1969*b*) and the montane bog species of *Metrosideros* (Sastrapadja and Lamoureux, 1969).

Vessels angular in transection have been considered primitive for dicotyledons by Frost (1930*b*). They probably are, but these relatively weak vessels are tolerable only by the simultaneous incidence of thick-walled tracheids, as in *Illicium* (plate 8-E). In fact, one can say that in woods with scalariform perforation plates, tracheids with wall thickness greater than that of vessels represent an almost inevitable component of division of labor between tracheids and vessel elements. One may hypothesize that in the primitive woods with scalariform perforation plates, the total utilization of both perforation and lateral wall pits as conductive areas is permitted by the mechanical compensation of thick-walled tracheids. Thus the thin-walled, heavily pitted vessel element does not develop improvement in mechanical characteristics, and does not develop the vessel strength characteristics of more specialized woods, where vessel walls evidently play an important mechanical role. Occurrence of negative pressures may be minimal in the primitive woods; mechanically weak vessel elements are thus not disadvantageous. As long as tracheids are the imperforate element in a primitive wood, they retain a conductive function, and thus retention of abundant pitting on the lateral walls of vessels facing them may be of selective value.

11. Relatively few vessels with scalariform perforation plates are exceptionally wide, and they tend to be narrower than in dicotyledons as a whole (figure 13-B). If the lateral walls are not strong, narrowness is of selective advantage (as with conifer tracheids), and only a strong selective pressure for wideness can result in marked widening. If a species experiences consistently low negative pressures in the xylem, there is not a strong selective pressure against wideness, but low negative pressures are closely tied to low conductive rates in many woody plants, so that narrow vessels would be still probable under these circumstances.

12. Frost (1930a) and Cheadle (1943) showed by statistical correlations a phylesis from vessels angular in transection to those circular in outline. What is the basis for this phylesis? Scalariform pitting conforms to the faces of an angular vessel, but if the vessels are circular in transection, alternate pits form a stronger pattern. The distribution of wall material between alternate pits represents an equidistant distribution of thickened areas, rather like the struts in a geodesic dome. In dicotyledons, this stronger pattern is also correlated with division

of labor. Imperforate elements become progressively less conductive and more mechanical (libriform fibers), which in turn minimizes conduction via the lateral walls of the vessels. The loss of all bars on the perforation plate is concurrent. With the loss of this impedance, the value of the imperforate elements for conduction, and of the lateral walls of the vessels for conduction, is much lowered.

Opposite pitting is a method of adding more lateral wall material (in contrast to scalariform pitting), perhaps with a lowered value of the lateral walls of the vessels for conduction. Frost (1930b) claims that progression toward alternate pitting on the lateral walls of vessels progresses more rapidly than perforation plate specialization. The selective value of stronger lateral walls may be the explanation. For example, in *Austrobaileya* (plate 11), perforation plates are scalariform, but lateral walls have alternate pitting (plate 11-B). This probably represents a compensatory strengthening of vessel walls with development of relatively wide vessels in this lianoid genus. Widening of vessels also means that the number of cells contacted by each vessel is more numerous. Paucity of cell contacts of narrow vessels was cited above as a reason for the correlation sometimes observed between narrowness of vessels, their angularity, and the tendency for lateral walls to bear scalariform pits.

13. If (see no. 2, above) long fusiform cambial initials yield long vessel elements with long overlap areas, they also provide long imperforate elements. Long imperforate elements are of value for their greater mechanical strength, as suggested in discussions of the conifer tracheid in chapter 7. Long imperforate elements are achieved with little intrusive growth if fusiform cambial initials are nearly as long as the imperforate-element length optimal for a given species. Therefore, intrusive capability of imperforate elements as they mature is not highly developed. More specialized dicotyledons have developed a much greater degree of intrusiveness in imperforate elements. A high degree of intrusiveness is likely correlated

with less pitting of imperforate elements, and therefore accompanies the division of labor which mechanical elements represent in specialized dicotyledon woods.

14. As is obvious from figure 11, a marked drop in the average length of tracheary elements—and therefore also fusiform cambial initials—takes place concomitantly with the origin of vessels. Division of labor between conductive cells and mechanical cells is an obvious reason for this. The great length of tracheids in a vesselless wood is concomitant with the great length of overlap areas, which provide greater conductivity by this area despite the inefficiency of the bordered pit. As soon as perforations are evolved, the value of a long end wall, and thereby part of the value of a long element, is diminished.

Long mechanical cells offer strength in proportion to their length by virtue of underlying cellulosic organization (see chapter 1). Why are not all tracheary elements in a vessel-bearing wood long, then? Long vessel elements are disadvantageous because their great length (together with width greater than that of a tracheid) makes them vulnerable to collapse when the water columns they contain are under tension. This vulnerability can be overcome by thickening the walls or by shortening the vessel elements, or both. Extremely long vessel-elements are least vulnerable to collapse when they are narrow, in which case their conductive function is negated because of increased friction.

However, mechanical elements in a vessel-bearing wood can be thick walled, because they are "released" from the broad-lumen configuration characteristic of the average coniferous tracheid. This "release" is obvious in primitive woods, and shows that the compensation (by wall thickness) for shortness of imperforate elements (as compared with those in vesselless woods) does occur. Tracheids in most primitive vessel-bearing woods have walls between 5 and 15 μ in thickness, as in *Illicium* (plate 8-E).

15. The division of labor in a primitive vessel-bearing wood

is different from the division of labor in a vesselless wood in terms of the proportion of mechanical and conductive elements. If one views transections of vesselless woods, either coniferous (plate 3) or dicotyledonous (plates 4, 5, 6, 7), one sees that the proportion of earlywood tracheids to latewood tracheids is at least 1:1. In many, the proportion of earlywood tracheids rises much higher, to about 9:1.

In a vessel-bearing wood, the proportion of conductive to mechanical cells can be reckoned as the number of vessels compared to the number of tracheids as seen in transection. If we view such transections (plate 9-E, for example), we find that the proportion of conductive to mechanical cells is reversed compared to that of a vesselless wood in which growth rings occur. Over 90 percent of the tracheary elements in a vessel-bearing wood are mechanical elements; 10 percent or fewer are vessel elements. Thus, the greater proportion of mechanical elements alone would compensate for their shortness compared with tracheids in a vesselless wood. Exceptions to this would occur only in scandent species, such as *Austrobaileya* (plate 11-A), where mechanical strength is of much less selective value (chapter 11). In this regard, it is interesting to note that no conifer or vesselless dicotyledon is scandent or lianoid in habit: the demands for rapid conduction in this growth form cannot be satisfied, evidently, by tracheidal systems. *Gnetum* (some species of which are lianoid) is, of course, a vessel-bearing gymnosperm.

VESSEL ORIGIN IN QUANTITATIVE TERMS

If one wishes to see modes of origin of vessels, the obvious place to look for exemplary conditions would be in a woody family in which both vesselless and vessel-bearing conditions occur. There is such a family: Chloranthaceae. *Sarcandra hainanensis* has relatively short (1,680 μ) tracheids for a vesselless plant (fig. 11), probably because it is a fruticose species. In the closely related *Chloranthus officinale*, there is an immediate drop in tracheary-element length (vessel elements, 800 μ; tracheids, 900 μ, according to Bailey and Tupper, 1918). Swamy (1953) gives for *"Chloranthus"* ($=$ all species he studied), average vessel-element length, 856 μ; tracheid length, 922 μ. The figures form a tracheid-to-vessel-element length ratio of 1.07. The ratio is also very low (1.05) in *Hedyosmum nutans*, another member of Chloranthaceae. One may note that in figures 12 and 14-A, the chloranthaceous species *Hedyosmum nutans* ("27"), *H. bonplandianum* ("26") and *Ascarina* sp. ("28") all approach the maximal tracheid-like limit expected of vessel-elements with primitive characteristics.

Also notable in this regard are *Aextoxicon punctatum* ("6") with a tracheid–vessel-element ratio of 1.18, *Sphenostemon lobospora* ("2") with a ratio of 1.04, and *Pentaphylax clethroides* ("24") with a ratio of 1.10. The tracheid–vessel-element length ratio for the 28 species studied as a whole is 1.36, a low ratio compared with that of more specialized dicotyledons (fig. 11). If one were to guess on the basis of those species with a low ratio, one would say that origin of vessels in primitive woody dicotyledons may have taken place in small trees, probably of the understory of forests in moist subtropical areas or tropical uplands, and that tracheary-element length following the origin of vessels was about 1,600 μ. *Sarcandra* and *Chloranthus* probably do not have a typical habit for primitive dicotyledons, and there is no reason why origin of vessels could not have taken place in several groups of vesselless dicotyledons—in fact, the circumstantial evidence of the highly diverse floral morphology and growth forms in groups with primitive vessel-bearing woods suggests polyphyletic origin of vessels in dicotyledons.

Lengths of vessel elements are much greater in the 28 species with primitive wood than in dicotyledons as a whole (fig. 13-A). In diameter, however, the 28 woods show a distribution curve (fig. 13-B) not markedly dissimilar to the dicotyledons as a whole. The curve shows fewer wide vessels in primitive woods, which would be expected on the basis of considerations above, and the abundance of narrow vessels in primitive woods seems a statement of their greater impedance to conductive flow. Selection for greater conductive efficiency is suggested by a family with relatively primitive characteristics according to phylogenists, Myristicaceae. Three species of that family, *Osteophloeum platyspermum*, *Iryanthera hostmanii*, and *Virola multicostata*, have relatively long vessel elements (1,181 μ, 1,185 μ, and 1,182 μ, respectively) with simple perforation plates and wide diameters (275 μ, 119 μ, and 142 μ, respectively).

Does tracheary-element length in the 28 species bear a relationship to plant size, as in gymnosperms and vesselless dicotyledons? Not a close one. The longest vessel elements in the group occur in *Strasburgeria robusta*, which is at most only a small tree, and I took the wood sample studied here from a tree less than 10 feet tall. Shortness of elements does characterize *Carpenteria californica*, a shrub. The species of *Eucryphia* are only small trees. The stem of *Pentaphylax clethroides* studied was relatively small, so that short vessel elements (ave., 766 μ) could have been expected on the basis of age-on-length curves (e.g., fig. 15). However, the stem of *Sphenostemon lobospora* (vessel elements, 1,590 μ) was equally small. I suspect that in vessel-bearing dicotyledons, the pattern of lengths is based on factors quite different from those in vesselless woods. Greater length of tracheids in a vesselless wood connotes both greater mechanical strength *and greater conductivity* (chapter 7). With origin of vessels, length of vessel elements does not connote greater conductivity, especially with simplification of the perforation plate. As mentioned earlier in this chapter, thickness of imperforate element walls and the proportion of imperforate elements to vessel elements provide factors quite different from those operative in a vesselless wood. Ray length, width, and ray cell wall thickness can contribute to the degree of mechanical strength in a wood. Thus, imperforate element length is not directly related to strength in those dicotyledons with vessels, and thus neither of the values of greater tracheid length operative in vesselless woods is operative in vessel-bearing woods.

Internode length may relate to tracheary-element length in monocotyledons, as discussed in chapter 8. However, internode length does not appear to be a significant factor in trachearly-element length in gymnosperms and dicotyledons. The example of *Echium pininana* (table 11) shows how internode length runs counter to tracheary-element length, for example.

GNETALES AS COMPARED TO PRIMITIVE
VESSEL-BEARING DICOTYLEDONS

Consideration of Gnetales is appropriate at this point because the woody genera, *Ephedra* and *Gnetum* resemble primitive vessel-bearing dicotyledons in their stage of release from the vesselless condition, although Gnetales are not related to dicotyledons, of course. A wood sample from a large plant of *Ephedra viridis* (C.B. Wolf 6971, RSA) had average tracheid length 709 μ, average vessel element length 677 μ (ratio: 1.05). In a tree specimen of *Gnetum gnemon* from New Guinea (Carlquist 1329, RSA), I found tracheids to average 1,650 μ in length, vessel elements 1,575 μ (ratio: 1.04). In a lianoid stem of *Gnetum indicum*, a greater divergence (which one would expect in a scandent habit, as suggested in chapter 11) was observed (tracheids, 1,372 μ; vessel elements, 1,164 μ; ratio: 1.18).

The reason for the low degree of divergence between vessel elements and tracheids in Gnetales is not retention of a perforation plate. Perforation plates are simple in the *Gnetum* species studied, but multiperforate in the *Ephedra*. Rather, the morphology of the tracheids appears to be the overriding factor here. Tracheids in Gnetales have numerous large circular bordered pits, virtually identical to those on the lateral walls of vessels. This is quite unlike vessel-bearing dicotyledons, where lateral wall pitting of the primitive vessel is already sharply different from pitting on imperforate elements. Tracheids in Gnetales have continued as conductive elements, have not assumed a prominent mechanical role, and division of labor is minimal; or one can say that the conductive capabilities have been improved much more than have the mechanical capabilities. In this connection, we may note that the two woody genera of Gnetales seem ecologically very restricted to opposite habitats. If the conductive system of the two genera is still "tracheid oriented," as it appears to be, they would tend to reflect restrictions similar to those of conifers. *Ephedra* is like a microphyllous desert conifer, such as *Juniperus*, with which *Ephedra* may be found growing in some localities. Likewise, *Gnetum* is broad-leaved and characteristic of humid tropical forest, as is *Podocarpus blumei* (which may be found growing with *Gnetum gnemon*).

There may seem to be an irony in the fact that *Ephedra* retains foraminate perforation plates (numerous circular perforations) in xeric habitats whereas *Gnetum* has predominantly simple perforation plates yet grows in mesic habitats. Several reasons can be advanced for this situation. First, conductive and transpiration rates in *Ephedra* are probably low, except during the short bursts of growth when water is available. The presence of scale-like leaves and the fact that stomata on stems are plugged (Wulff, 1898) suggest slow transpiration rates. If conductive rates are quite low, scalariform plates are of no great advantage compared with simple perforation plates (and the pore areas of *Ephedra* plates are comparable in total area per plate to simple plates). The wide earlywood vessels in *Ephedra* are still much narrower than earlywood vessels in mesic dicotyledons, a fact true of desert shrubs at large. The narrowness of vessels and the presence of perforation plates suggest a second possibility. Although sap tensions have not, to my knowledge, been measured for *Ephedra*, one can assume on the basis of other desert shrubs

(Scholander, Hammel, et al., 1965) that negative pressures are high. If so, the strength offered by narrow vessels and by presence of foraminate plates would prevent collapse of vessels.

Relatively lower tensions, yet probably rapid conductive rates (especially in lianoid species) probably occur in *Gnetum*. Therefore, simple perforation plates are advantageous. Vessels in *Gnetum* are also rather wide compared with those of *Ephedra*, so that as mentioned before, presence of a perforation plate would negate the advantage gained by a wide vessel. In the lianoid species, accelerated loss of perforation plates would be expected, as mentioned earlier in this chapter.

Simple perforation plates in *Welwitschia* may relate by way of compensation to the limited quantity of vascular tissue in that plant. Perhaps high conductive rates occur when strobilar shoots are in a period of rapid growth, and lend an advantage to simple perforation plates.

ORGANOGRAPHIC ORIGIN OF VESSELS IN DICOTYLEDONS

Bailey (1944b) has hypothesized, on the basis of comparative studies, that vessels in dicotyledons originated in the secondary xylem of roots and stems simultaneously. In view of the present considerations, we can see why this is probable. In monocotyledons, roots are exclusively adventitious, except for the short-lived primary root of the seedling, and provide a differential between roots and stems. Dicotyledons, on the contrary, have an inescapable continuum between secondary xylem of stems and roots. Any root system that develops secondary xylem cannot be isolated from stem secondary xylem. Because vessels (a series of interconnected vessel elements) may extend for an indefinite vertical distance in the plant, occurrence of vessels in secondary xylem of stems, without a continuation of these numerous vessels into the root, is unlikely morphologically—as well as physiologically. There would be virtually unmanageable conductive problems in woody vessel-bearing plants if root vessels did not interconnect with stem vessels. Also, the release, by virtue of vessel origin, into conductive and mechanical systems in the secondary xylem is just as potentially advantageous in roots as it is in stems. I would guess that these considerations would apply to *Gnetum* and *Ephedra* also, and that vessels originated simultaneously in stems and roots of these two genera.

Bailey (1944*b*) has also hypothesized that in dicotyledons, vessels have phylogenetically been extended into primary xylem after their origin in secondary xylem. Bierhorst and Zamora (1965) support this concept in general, but note some refinements. Their list of species in which simple perforation plates characterize the sequence from protoxylem to metaxylem (but with occasional scalariform perforation plates on apically elongate tracheary elements) can be said to consist of species in which the simple perforation plate pattern and other specialized characters are well established in the secondary xylem. Thus, there may not be any overriding reason other than the eventual (but delayed) inevitability that patterns well fixed genetically in secondary xylem eventually will occur in primary xylem as well. The data of Bierhorst (1960) on primary xylem of Gnetales suggest that as in dicotyledons, vessels may have occurred in secondary xylem before they have been introduced, evolutionarily, into primary xylem.

As is well known (Bierhorst, 1960), circular bordered pits can be found interpolated into helices and other wall thickenings of primary xylem-elements in gymnosperms. This suggests that, paralleling the occurrence of vessels in dicotyledons, pitting patterns characteristic of secondary xylem tracheids in gymnosperms have perhaps been introduced evolutionarily into primary xylem secondarily.

Bailey's concept of progression of secondary xylem characteristics into the primary xylem is clearly demonstrated by Bierhorst and Zamora (1965), but they are in error in saying (p. 701) that my theory of paedomorphosis (1962) is a restatement of this concept. In fact, it is a reverse tendency: extension of primary xylem characteristics into the secondary xylem, a tendency that occurs only under special circumstances and in certain growth forms and is not characteristic of dicotyledons in general.

II.

Specialization in Dicotyledonous Wood

ROLE OF VESSEL-ELEMENT DIMENSIONS AND MORPHOLOGY

The transition from scalariform to simple perforation plate has undoubtedly occurred numerous times independently in dicotyledons, and may rightly be considered polyphyletic. One reason for this statement is the systematic distribution of perforation plate types. Numerous families contain some genera with scalariform perforation plates, and some genera with simple perforation plates. This is also true of species within genera.

One advantage of attainment of the simple perforation plate is the gain in capability for conducting larger volumes of water per unit of time. The presence of a few vestigial bars on a perforation plate offers only a little impedance to water flow and only a minor amount of strengthening. With a small amount of either positive or negative selective value, the last vestiges of the scalariform condition may not be expected to disappear slowly. Nevertheless, the advantages of the simple or near simple perforation plate appear so great in terms of the capability for phylads to radiate into new habitats that passage over the threshold to simple or near-simple plates may be considered basic to the diversification of the major groups of dicotyledons. There are few genera or species with a small but constant number of bars per perforation plate unless, as in Rhizophoraceae, those bars are wide and appear to have a definite mechanical significance. In such families as Dichapetalaceae and Myricaceae, scalariform perforation plates occur occasionally along with simple perforation plates in a given section of wood.

Vessel-Element Length

With elimination of the scalariform perforation plate during phylesis, vessel-element length continues to shorten. Because vessel-element length is roughly the same as fusiform cambial initial length, phyletic decrease in initial length continues to take place well after the simple perforation plate has been attained. This situation can be highlighted with an example, table 10.

TABLE 10

Vessel-Element Length for Species in Which
Imperforate Elements Average about 1600 μ

Species	Perforation Plate	Vessel-Element Length, Average
Laurelia novae-zelandiae	more than 20 bars	1400 μ
Gordonia lasiantha	more than 20 bars	1300 μ
Cassipourea elliptica	2–5 bars	1000 μ
Magnolia acuminata	fewer than 20 bars	800 μ
Grewia stylocarpa	simple	700 μ
Platanus occidentalis	simple	500 μ
Grevillea robusta	simple	400 μ
Fremontodendron californicum ..	simple	200 μ

DATA from Bailey and Tupper, 1918.

The species of table 10 were selected because they all have a roughly identical imperforate length, yet a wide range of vessel-element lengths. These are all trees except for *Fremontodendron californicum* and perhaps *Grewia stylocarpa*, which are large shrubs.

In attempting to explain the progressive shortening in vessel elements in dicotyledons, two features must be taken into account: the shortening of fusiform cambial initials and the increasing intrusiveness of imperforate elements as they mature. The apparent value of shorter vessel elements is their structural resistance to strong negative pressures in the water columns of the xylem—a feature that was cited for explaining shortness of tracheids in some conifers (chapter 7). The following points can be cited:

1. Webber (1936), in a survey of desert and chaparral shrubs, found that vessel elements in these shrubs are much shorter and narrower than in dicotyledons at large. Examples of desert shrubs of which wood is illustrated here include *Larrea divaricata* (plate 15-A), which has vessel elements averaging 123 μ in length in this sample, and *Artemisia arbuscula*, in which vessel elements average 116 μ (plate 15-B). These species also have narrow vessels: 39 μ and 26 μ average diameter, respectively. *Larrea tridentata*, according to data by Scholander, Hammel, et al. (1965: *Larrea* cited as "creosote bush") has exceptionally high negative pressures in xylem: −60 to −80 atmospheres. There seems to be a clear correlation between dimensions of elements and the high negative pressures. Novruzova (1968) found the shortest vessel elements in dry sites.

2. In table 10, the species from the most xeromorphic habitat is the species with the shortest vessel elements, *Fremontodendron californicum*.

3. In my (1966a) survey of wood anatomy in Asteraceae, data on all species studied in the family were averaged according to habitat categories: "mesic," "dry," and "desert." Because of the large number of species involved, the data seem reliable in showing distinct decrease in both vessel diameter and vessel-element length with increasing aridity. The only member of Asteraceae in the survey by Scholander, Hammel, et al. (1965) was *Encelia* (presumably *E. farinosa*), in which tensions just below −30 atmospheres were measured. This may seem not striking compared with the negative pres-

sure in *Larrea*, but it is at least as great as the negative pressure at very high levels in a coast redwood (between −15 and −17 atmospheres at 82 m.) according to those authors.

4. Stem parasites (epiparasites) such as mistletoes have higher negative pressures than their hosts in each of several cases examined by Scholander, Hammel, et al. (1965). A mistletoe illustrated here (*Phoradendron flavescens macrophylla*, plate 14-A,B) shows exceptionally short, narrow vessel elements. I made wood macerations of three mistletoes and their hosts, and found vessel elements shorter in the parasite than in the host in each instance.

5. The narrow, short vessel elements of species cited above are also exceptionally thick walled. This can be seen in *Phoradendron flavescens macrophylla* (plate 14-A), *Larrea divaricata* (plate 15-A), and *Artemisia arbuscula*. One might note, by contrast, the thin walls of vessels of *Illicium cambodianum*, shown at relatively higher magnification in plate 8-E. I have mentioned in chapter 10 that mangroves experience great negative pressures: between −40 and −60 atmospheres (Scholander, Hammel, et al., 1965). Thick-walled vessels would be expected in mangroves, and this is true, as shown for *Ceriops tagel* (plate 14-C). Exceptionally wide vessels of vines, which can experience appreciable negative pressures, ought to be thick walled for structural reasons; this appears true, as shown for *Doxantha unguis-cati* (plate 12-C). In mesomorphic plants, if vessels are narrower, thinner walls ought to suffice, for similar structural reasons. This is true in Hawaiian *Metrosideros*. Sastrapadja and Lamoureux (1969) have shown that in upper portions of trees, vessels are narrower and shorter and also thinner walled than those of basal portions of trees. The smaller dimensions in vessels of upper portions of a tree is interesting, for it would accord with the Scholander findings that tensions in water columns are greater at upper levels than at lower levels in trees.

6. Borders on simple perforation plates are definitely a potential strengthening feature. These borders are narrower

than the vessel itself, and bear no pits. They thus appear like constrictions at the end of each vessel element, and because they are non-pitted they appear to be strong rings. The shorter a vessel element, the more such rings per unit length of vessel there would be in a wood. If these borders do lend strength to the vessel, shorter, narrower vessel elements would be more resistant to negative pressures.

7. Very short fusiform cambial initials occur in woods with storied cambia (Beijer, 1927). These very short fusiform cambial initials have minimal overlap areas and are the most likely to produce, as derivatives, vessel elements with transverse perforation plates. If both shortness and transverse perforation plates indicate strength, and if desert shrubs might be expected to have vessels resistant to negative pressure, the percentage of woods with storied structure ought to be highest in desert areas. This is true in Asteraceae (Carlquist, 1966a).

8. Alternate lateral wall pitting offers maximal strength to lateral walls of vessels, as discussed in chapter 10. Alternate pitting is associated with simple perforation plates and short vessel elements (Frost, 1931), especially minute pitting (Metcalfe and Chalk, 1950, p. xlvi). With the progressive specialization in dicotyledon woods, the borders on imperforate elements are reduced, so that specialized woods have libriform fibers rather than tracheids or fiber-tracheids with minimal conductive ability compared to tracheids; lateral wall pitting to libriform fibers from vessels becomes of minimal importance. Thus reduction of the total pit area in the lateral walls of vessels can be reduced with phylogenetic specialization, but only in the case of contacts other than vessel-parenchyma. Metcalfe and Chalk (1950, p. xlv) demonstrate that pit borders in imperforate elements decrease markedly with vessel specialization (shorter vessel elements with transverse end walls), thus resulting in a higher proportion of libriform fibers among species with more specialized vessel elements. This correlation by Metcalfe and Chalk is a particularly interesting one, because it is a measure of progressive division of labor

between mechanical and conductive systems in dicotyledonous wood.

9. Wood of roots, when compared with that of stems, is very informative. Few workers have consistently compared stem and root wood for ·a series of species. Patel (1965) studied several common North Temperate tree species in this regard. Henrickson (1968) offered data on root and stem wood for each species of Fouquicriaceae. Ingle and Dadswell (1953) compared root and stem wood in a mangrove, *Alstonia spathulata*. A few comparisons of roots and stems may be found in Gibson's (1973) study of wood anatomy of cereoid cacti. Table 11 offers data for wood of roots and for various levels of the stem of *Echium pininana*.

With some exceptions, the results of the above studies show that vessel elements tend to be wider and longer in roots than in stems in any given species. If shorter, narrower vessel elements connote greater resistance to negative pressure of water columns, the longer, wider vessels of roots would suggest that roots experience less water tension than do stems. In fact, this is what the results of Scholander, Hammel, et al. (1965) show. These results are consistent not only with wood anatomy, but with the Dixon-Askenasy theory of ascent of sap. As should be expected, the tendency for tracheids in conifers to be longer and wider in roots than in stems (chapter 7) accords with the data from dicotyledons.

10. If higher rainfall or relatively uniform rates of transpiration in mesic situations provide lower tensions in the conductive systems of woods, woods in these situations would be expected to have longer, wider vessel elements. This is the case in Hawaiian *Euphorbia* species (Carlquist, 1970b); rain forest species are almost certainly derived from lowland species. In this case, there has been a "release" to wider, longer vessel elements. Although the overall trend in dicotyledons may be to shorter vessel elements (probably because xeric conditions are less "saturated" with species than are the mesic areas), occasional opportunities offer a reversal, and the dis-

harmonic wet forests of the Hawaiian Islands present such
an opportunity. Stages along the transect from dry lowland
to wet montane sites show intermediacy in vessel dimensions
in *Euphorbia*. Wider vessel elements would be of positive
value in this "release" into more mesic conditions because
they offer less friction in conduction, but the fact that less
strong tensions would exist in xylem of wet forest species may
be the overriding factor. Longer vessel elements may not have
a positive selective value by themselves in more mesic sites
in *Euphorbia*, but they are at least the byproducts of longer
fusiform cambial initials. Longer cambial initials also produce
longer imperforate elements, which might have a positive
value in the woods of the more arborescent growth forms in
the wet forest. Which of these two factors is more important
in the evolution of the longer cambial initials is not easy
to discern.

However, other similar cases of adaptation to more mesic
conditions on islands can be cited. Progression into wet forest
is accompanied by evolution of wider, longer vessel elements
in the Macaronesian species of *Euphorbia* (Carlquist, 1970*b*),
the Macaronesian species of *Sonchus* (Carlquist, 1974) and
Echium (Carlquist, 1970*a*), and in the Hawaiian *Dubautia–
Argyroxiphium–Wilkesia* complex (Carlquist, 1974).

In Asteraceae as a whole, the same trends hold (Carlquist,
1966*a*). Tree Asteraceae are probably secondary entrants into
mesic sites from phylads with more xeric preferences. Even if
mesomorphy does not provide the basic selective factor for
length of elements in these, one cannot deny that greater
diameter of vessel elements is a "release" related to mesomorphy, for vessel diameter evolves independently of vessel-
element length.

Longer, wider vessel elements in tropical species compared
to those of temperate species can be cited for *Nothofagus*
(Dadswell and Ingle, 1954) and *Prunus* (Baas, 1973). In
these instances, the correlation with climate and physiology
certainly holds, although no opinion has been given as to

whether the evolution in these two genera has been from tropical to temperate or the reverse.

11. Vines and lianas have notably wide, short, thick-walled vessel-elements (plate 12-C,D; fig. 14-B). The wideness may be considered an adaptation to rapid rates of flow, rates demonstrated by Scholander, Hemmingsen and Garey (1961). One would judge that lianas and vines do not experience strong negative tensions in xylem, if the data of Scholander, Love and Kenwisher (1955) are indicative, and this would permit wideness. However, even with moderate negative tensions wide vessel elements are vulnerable to collapse when long (this is less true of short vessel elements), if the empirical data provided by the dimensions of vessel elements of vines and lianas are indicative. The thickness of vessel walls in vines has been mentioned earlier, and in chapter 10 the suggestion was advanced that *Austrobaileya* has alternate pits on lateral walls in response to the selective value of strong walls in vessel elements of a scandent species.

12. The phyletic shift to wider, shorter vessel elements to accommodate a more rapid rate of flow in vines has a parallel in the primary xylem of dicotyledons, as well as other vascular plants. In the sequence from protoxylem to metaxylem, successively wider elements are formed (fig. 16). There is a concomitant shift to shorter elements. Wider elements represent a capability for greater volume of flow. With wider elements, there must be shortness to counter the vulnerability of wider elements to collapse under tension.

13. There is no straight-line correlation between diameter and length in primitive dicotyledons taken as individual species (fig. 14-A). This holds true for specialized dicotyledon woods also. Control of vessel wideness is, of course, independent of the fusiform cambial initial control of length. If one groups species with, say, short vessel elements (e.g. desert Asteraceae), one finds that they have, as a group, relatively narrow average vessel diameter. The structural advantage conferred by short, narrow vessels that permits their functioning without collapse

under high water tension is not disadvantageous if, in any given species, wide vessel elements of the same length can be formed seasonally. Thus, desert shrubs can form wide vessels during moist conditions, vessels that permit greater flow. The narrow latewood vessels that resist higher tension during the dry portion of the year suffice for conducting smaller water volumes available then.

The model for desert shrubs is more conspicuously realized in ring-porous, temperate-zone trees, and the same factors apply. Although we have less data on alpine shrubs, such as *Loricaria thuyoides* (plate 15-C,D), they probably follow a pattern not unlike that of desert shrubs.

14. Decrease in vessel-element length (and therefore also imperforate-element length) occurs ontogenetically in the secondary xylem of some dicotyledons (fig. 17). This connotes a decrease in the mechanical strength for some of these— particularly when vessel-element diameter stays constant. However, in the case of the stem of *Aster spinosus* in figure 17, not only do elements become shorter, vessels become markedly narrower toward the periphery of the stem as growth ceases. This is true of stems of annuals that terminate growth under water stress conditions. In these, there is no value for mechanical strength at the end of the growing season with respect to imperforate elements, but there is a strong value for the resistance to high tensions in xylem, and the formation of very narrow, short vessel elements can have no other explanation.

Figure 14. Graphs comparing: A, vessel diameter to vessel-element length; and B, vessel diameter to number of vessels per square millimeter. The species in A are those listed in the legend for fig. 12. The species in B (represented by dots) on which the straight-line curve has been based, are the 28 species of fig. 12, plus species from Myristicaceae, Monimiaceae, Icacinaceae, and Rhizophoraceae. Also in B, species representing various growth forms have been plotted: A = anomalous (woods with successive cambia); D = desert shrubs; S = stem succulents; V = vines and lianas; species from the categories plotted represent only a selection of the species studied in each growth form.

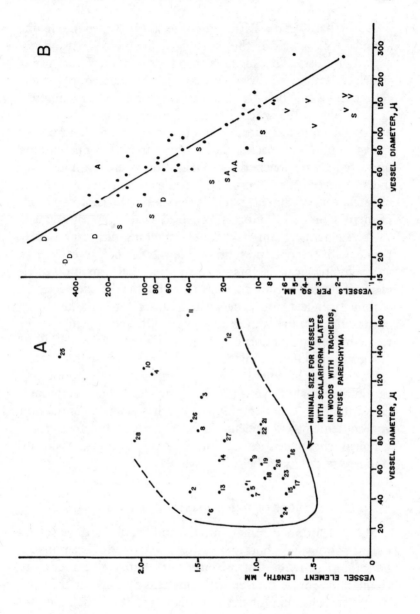

15. In hardwoods, Ritter and Fleck (1926) found lignin abundant in vessels of the earlywood; lignin in latewood occurred on fibers. This suggests that the strength of earlywood vessels is enhanced by their chemical composition, which compensates for the potential weakness of a greater diameter.

16. Bailey (1957) cautioned against the use of vessel-element length too literally as an indicator of phylogenetic specialization in a species. Bailey understood that fluctuations unrelated to phylogenetic specialization occur, although he did not investigate their causes. Some workers have, in fact, claimed that within a genus, a species with shorter vessel-element length was more specialized than one with longer vessel elements. I am deliberately avoiding citing particular papers with this interpretation. One should look to ecological and physiological factors for variation patterns in vessel-element length within a genus, as well as within a species or within a single plant. Unexpected causes for variations may occur. Higher polyploidy is correlated with longer, wider vessel elements and longer libriform fibers in *Parthenium argentatum* (Swamy and Govindarajalu, 1957). Discussion of the significance of vessel-element length is not complete without also cautioning that there may be instances where tracheary-element length is determined not by physiology of conduction and consequent vessel-element characteristics, but by the mechanical characteristics of imperforate elements. Such instances are discussed below.

ROLE OF IMPERFORATE ELEMENTS

Short fusiform cambial initials that yield short vessel elements also yield short derivatives destined to become imperforate elements. Intrusiveness of maturing imperforate elements becomes greater with phyletic advance in dicotyledons, compensating for the short length of the initial. Despite this tendency, there are probably upper limits on the degree of intrusiveness, judging from imperforate-element lengths in specialized dicotyledons. If longer imperforate elements have a selective advantage in any species, longer fusiform cambial initials may be advantageous, provided that there is no conflict with the optimal length of vessel elements as described in the preceding discussions.

Attempting to isolate instances in which the length of imperforate elements is of prime importance in determining the selective value of fusiform cambial initial length is not easy. For example, plants in which the highest negative pressures tend to occur in xylem are desert shrubs (Scholander, Hammel, et al., 1965). Short vessel elements are to be expected in these, for reasons stated earlier. However, shorter imperforate elements are not incongruent with the shrubby habit, as in shrubby conifers. At the other extreme, if a phylad adapts to more mesic conditions, as in the case of the Hawaiian Euphorbias cited above, longer vessel elements are not disadvantageous with much more moderate negative pressures, and longer imperforate elements are probably advantageous. Because the maximal limits of plant size are roughly correlated, in woody groups, with the degree of water availability, there is little conflict between selective factors affecting length in the conductive and mechanical systems of the wood.

There are some instances in which the length or other histological features of imperforate elements seem of prime significance in evolutionary patterns of secondary xylem that can be cited, however.

1. *Echium pininana* (Boraginaceae) is a single-stemmed monocarpic rosette "tree" (table 11). In its native habitat (the laurel forests of La Palma, Canary Islands), the stem of *Echium pininana* bears a rosette of leaves until it is about 8 feet tall. The inflorescence (above "level 6" in table 11) extends the plant height to about 12 to 15 feet. The woody cylinder is thick (about 4 cm. in diameter) at the base of the plant, with a small pith. Upward, the pith becomes wider, the woody cylinder thinner, as is typical for dicotyledonous rosette trees.

The data of table 11 show that vessel diameter from the base of the plant to the top remains approximately constant. However, vessel-element length increases markedly at higher levels in the plant. If ecological factors affected vessel-element length, and the length of vessel elements were the selective factor of importance, this pattern would make no sense: we would have vessel elements characteristic of a xerophyte at the base, but more like those of a mesophyte in the inflorescence. The number of vessels per square mm. does fluctuate, but not markedly, in the stem. Thus, the only significant pattern with relation to the habit of the plant is provided by the length of the libriform fibers in the stem. The longer fibers at higher levels would compensate for the thinness of the woody cylinder, so that strength is maintained. The factor of internode length does not affect element length. Internodes are longest a short distance above the base of the plant, as it grows rapidly from the forest floor, then shorter as the rosette reaches better-illuminated levels, then longer in the inflorescences. Yet

TABLE 11

Quantitative Xylem Characteristics at Various Levels in a Single Plant of *Echium pininana* (Carlquist 2730, RSA). The Plant Was Cut into Lengths of About 18 Inches, so that "Level 3" Would Correspond to Approximately Four and One-Half Feet Above Ground Level

Portion	Vessel-Element Length (Average)	Libriform Fiber Length (Average)	Vessels per Group (Average)	Vessel Diameter (Average)	No. of Vessels per Sq. mm. (Average)
Root	268 μ	380 μ	1.5	127 μ	13
Base	233 μ	296 μ	1.7	81 μ	26
Level 3	291 μ	612 μ	1.8	94 μ	35
Level 6	322 μ	644 μ	2.1	86 μ	43
Level 10	576 μ	1030 μ	2.2	87 μ	51

vessel-element and imperforate-element length both consistently lengthen upwardly from the base of the plant.

With respect to vessels, one may note some significance in the number of vessels per unit transection. The increasing upward density of vessels in the plant compensates for the progressive narrowing of the xylem cylinder. In the root, wide vessels, slightly longer than those at the base, would correspond to tendencies in all woody plants: wider, longer tracheary elements because xylem tensions are lower in roots than in stems.

The pattern of the libriform fibers in *Echium pininana* is not unique. It applies to the species of woody lobelioids as well (Carlquist, 1969*b*). Other examples could be cited.

2. In particular genera where habit varies, and where differences among species cannot be ascribed to appreciable differences in phylogenetic level, habit may be related to imperforate-element length. Cacti would be a good example of this, because the non-leafy cacti ought to show minimal differences in transpiration and conductive rates among species on account of succulence. In the cereoid cacti, Gibson (1973) showed that tallness of the growth form is exactly correlated with length of the tracheary-elements. The length of libriform fibers appears to me more significant than the length of vessel elements. Indeed, the smallest globose cereoids lack libriform fibers altogether, and have only vessel elements and vascular tracheids. Lack of libriform fibers in secondary xylem of these obviously connotes complete relaxation of selection for mechanical strength. Even among species in genera, correlations between habit and fiber length are close: "within individual genera, such as *Lemaireocereus*, *Myrtillocactus*, and *Trichocereus*, degree of erectness correlates closely with average fiber length" (Gibson, 1973).

Succulents and "tank trees" escape from the high water tension patterns of desert shrubs. Whereas measurements by Scholander, Hammel, et al. (1965) show tensions of −20 atmospheres for shrubs of desert washes and from −30 to −80 atmospheres for shrubs of desert flats and hills, their measurements for the succulent trees *Idria, Fouquieria, Bursera,* and *Opuntia* range between −6.1 and −16.7 atmospheres—tensions comparable to those of mesic forest plants. Scholander, Hammel, et al. conclude, "like cacti, these plants avoid excessive tension by storing water." This explains why vessel-element length does not evolve with relation to external ecology (the dryness of deserts) in cacti. The longest imperforate elements (and vessel elements) are in such arborescent cacti as *Carnegiea gigantea, Cephalocereus brooksianus, C. polygonus, Cereus* cf. *dayami, C. jamacaru, Lemaireocereus godingianus, Pachycereus pecten-aboriginum,* and *P. pringlei* (Gibson, 1973). The fact that the mechanical strength of vessels to resist negative pressures in xylem is minimal is shown by the large pits on the vessel walls of cacti.

3. In stem succulents at large, such as *Talinum guadalupense, Brighamia insignis* (Carlquist, 1962), and *Crassula argentea* (plate 13-C,D) where mechanical strength is at a minimum, fibers are lacking. This can be seen in herbaceous stems of many dicotyledons. For example, in *Solidago spathulata,* the outermost portions of the secondary xylem show substitution of thin-walled parenchyma for libriform fibers. Interestingly, all stem succulents and succulent rosette trees surveyed in an earlier paper (Carlquist, 1962), like the cacti surveyed by Gibson (1973) show widening of pits in vessels, suggesting a loss of mechanical strength. The most extreme examples of this loss are: the small globose cereoid cacti, in which the only mechanical strengthening of vessels and vascular tracheids occurs in the form of widely spaced helices, mimicking primary xylem tracheary elements (Gibson, 1973); and in many Crassulaceae, with similar xylem configuration (plate 13-C,D). In these two cases, presence of helical and annular-like wall thickenings in tracheary elements, together with lack of fibers, may permit stems to accommodate a shrinkage of volume during drought without breakakge of

the conducting cells. One may note in passing that the primary xylem elements where large, "gaping" pit areas occur on walls in the survey of Bierhorst and Zamora (1965) represent instances where the mechanical strength of vessels is at a minimum. Such pitting occurs in elements of herbaceous species, or in certain monocotyledons where one can hypothesize that mechanical strength is borne entirely by bundle sheath fibers, and that the strength of tracheary-element walls is insignificant, and that tracheary elements are not, under natural conditions, subjected to high negative pressures.

4. Rayless woods often represent herbaceous species in which there is phyletic increase in plant size or woodiness (*Plantago princeps*, *Leptodactylon californicum*, *Viola tracheliifolia*) as I have hypothesized (1970c). The operative mechanism behind occurrence of raylessness can now be expressed as a temporary (or permanent, if plant size is limited and rays never develop) sacrifice in ray tissue in favor of mechanical strength. Characteristics of vessel elements in rayless woods are entirely unrelated to the presence of rays.

Aeonium arboreum (plate 13-A,B) and *Crassula argentea* (plate 13-C,D) can be presented as a pair in this regard. *Crassula argentea* has wide rays and abundant axial parenchyma, both non-lignified, and no libriform fibers in secondary xylem. The size of the plant is probably finite—rarely exceeding 1 m.—but mechanical support is achieved by the great width of stems, which consists mostly of cortical and pith parenchyma. Vessel elements have extraordinarily wide lateral wall pits, so wide that they are present in the form of annular thickenings rather than as pits.

The same type of pitting on vessel elements can be seen in *Aeonium arboreum*. Vessel groups are surrounded by thin-walled axial parenchyma. However, the remainder of the xylem is converted to libriform fibers. The potential ray areas are identical to fascicular areas in fibers. *Aeonium arboreum* is a succulent in which mechanical strength has been increased by raylessness. Many species of *Aeonium* and allied Crassulaceae are rosette herbs. *Aeonium arboreum* in this case represents increase in length of the stem so that something resembling a shrub, up to 1 m. tall, is produced. Rays apparently never develop in *Aeonium arboreum*. Lack of ray parenchyma is apparently ultimately a limitation. Rather than increasing indefinitely in height, stems produce adventitious roots abundantly and often eventually fall over, re-rooting the upper portions of a stem and thereby "short-circuiting" the older stems in which the limitations of the rayless condition, for whatever physiological reasons, are overcome. However, the pertinent point in regard to libriform fibers is that by the conversion of virtually all xylem parenchyma to fibers, *Aeonium arboreum* and other species of *Aeonium* have achieved increased height.

Raylessness is informative because it shows that alteration in the proportion of mechanical to conductive tissue can permit a shift in habit.

5. Intrusiveness of fusiform cambial initial derivatives destined to become imperforate elements increases phyletically (fig. 15). The 28 species with primitive wood have an average tracheid to vessel-element length ratio of 1.39. Woods with simple perforation plates and libriform fibers have a ratio of 2.60 (fig. 11). Some families in this category have a very high ratio. For example, Ulmaceae range as high as 9.50 (*Zelkova americana*). Other families with high ratios include Amaranthaceae and Nyctaginaceae (which have very short vessel elements); the malvalean families Bombacaceae, Malvaceae, and Sterculiaceae; and Moraceae, Urticaceae, and Proteaceae. These ratios cannot be used as measuring-sticks of phylogenetic specialization, although relatively low and high ratios are indicative.

Storied woods present a special problem. Because storied woods are those that show intrusiveness of imperforate elements, but to a uniform degree, we might tend to think that their imperforate-to-perforate-element length ratio is relatively low, and that these therefore ought to be regarded as

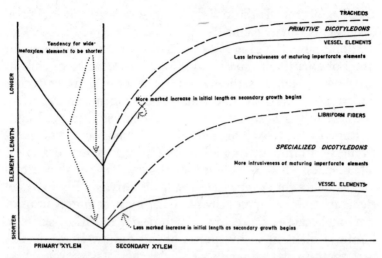

Figure 15. Age-on-length curves for dicotyledons with primitive woods; and dicotyledons with specialized woods. The curve for primitive woods corresponds to *Hedyosmum bonplandianum* (Chloranthaceae). The curve for specialized woods corresponds to *Carya ovata* (Juglandaceae).

primitive, which they definitely are not. In fact, even storied woods have a ratio of at least 2.0. However, storying of imperforate elements is not an accurate indication of a storied cambium. For example, *Fremontodendron californicum* (Sterculiaceae) has very long imperforate elements (ca. 1,600 μ) but short vessel elements (ca. 200 μ). Obviously, the imperforate elements show great intrusiveness after their derivation from the fusiform cambial initials, which may be assumed to be about as long as vessel elements. With such great intrusiveness, variation in the length of imperforate elements might be expected, and does, in fact, exist. Consequently, no storying appears in the libriform fibers, but the cambium is, nevertheless, storied. This can be seen indirectly in the fact that vessel elements and axial parenchyma conform to a storied pattern. One wonders how many woods with storied cambia have been overlooked because of a factor such as this. Dicotyledons with a storied wood evident in imperforate elements are probably mostly those with a moderate degree of intrusiveness of imperforate elements.

If one plots the range in the lengths of vessel elements in wood of a particular species against the percentage of cells in size classes, and does the same for libriform fibers (or other imperforate cells), one finds that vessel elements are relatively uniform in length, whereas there is a broad range in the length of imperforate cells. This indicates a varying degree of intrusiveness within a cell population, and only a low degree of intrusiveness; therefore, short imperforate elements show storying. My data for Asteraceae show that storying only in vessel elements and axial parenchyma is the most common form of storying because of a relatively high degree of intrusiveness of libriform fibers in many species of this family. If the species that show storying in imperforate elements have very short fusiform cambial initials (as they do, on the average) there is an explanation for existence of storying. The longitudinal radial division of fusiform cambial initials of storied cambia (the divisions that increase the girth of the cambium) would be expected to occur most readily in short cambial cells, and this expectation is, in fact, fulfilled.

6. In *Hibiscus*, herbaceous species have relatively long tracheary elements compared to woody species (Walsh, 1974). Longer elements—particularly imperforate elements—may compensate for a limited quantity of xylem in providing mechanical support in the herbaceous species. Similar considerations probably apply to the patterns reported by Cumbie and Mertz (1962) in *Sophora* and Anderson (1972) in *Bigelowia*.

7. Imperforate elements that are short can nevertheless give great strength to a wood by virtue of their relative abundance, wall thickness, and the nature of their wall chemistry. *Guaiacum officinale* (Zygophyllaceae) is a tree, yet its vessel elements are notably short (ca. 100 μ), and imperforate elements average only about 500 μ (Bailey and Tupper, 1918). However, *Guaiacum* has exceptionally thick-walled libriform fibers. This is also true in *Stenopadus*, a tree genus of Astera-

ceae (Carlquist, 1957). Wall thickness of imperforate elements could also explain the strength of woods of notably tall trees such as Ebenaceae and Fabaceae.

Sastrapadja and Lamoureux (1969) have shown in individuals of Hawaiian *Metrosideros* that wall thickness of fibertracheids is greater in upper portions of trees than in lower parts. Wall thickness also increases from inner to outer portions of a trunk. Greater wall thickness of fiber-tracheids may compensate for a lesser amount of secondary xylem in upper branches. This is almost certainly true in *Cyanea leptostegia* (Carlquist, 1969b). Greater wall thickness in outer portions of a trunk suggests that the mechanical value of xylem does not decrease with age.

8. The age-on-length curves for typical woody dicotyledons (fig. 15) show a sharp upswing in the length of tracheary elements after the beginning of secondary growth, either for primitive woods or specialized ones. The fact that any increase occurs can be related to the increasing mechanical strength produced by the greater length of imperforate elements. If mechanical strength were of no selective value as secondary xylem accumulates, one would expect no increase in length, or even decrease in length. In fact, this is what occurs in growth forms that show little stress on mechanical strength, and is related to paedomorphosis (Carlquist, 1962; see also fig. 17 and discussion later in this chapter).

The fact that the curve for libriform fibers (fig. 15) diverges from the curve for vessel elements in specialized dicotyledon woods has several interesting implications. The libriform fiber curve resembles closely curves for coniferous woods (fig. 7; *Pseudotsuga* curve in fig. 17-A) as well as the curves for primitive dicotyledonous woods (fig. 15, above) suggests that libriform fibers are following patterns dictated by the value of greater length for mechanical strength (for discussions and data supporting length as an indicator of strength, see chapters 1 and 7). Meanwhile, the curve for vessel-element length in specialized dicotyledons (fig. 15) shows a tendency toward shorter vessel elements, which would be stronger where resistance to negative pressures in water columns is concerned. Thus, the divergence between the libriform fiber and vessel-element curves is a measure of division of labor between the mechanical and conductive systems in more specialized dicotyledonous woods.

ROLE OF HELICAL THICKENINGS IN WOODS

Helical thickenings in vessels and other tracheary elements in secondary xylem—also called "spirals" and "tertiary helical thickenings"—do not show a wholly consistent pattern of occurrence in dicotyledonous woods. Baas (1973) cited data in terms of floras, based on data from Greguss (1959), Janssonius (1906–1936), Kanehira (1921a, 1921b, 1924) and Record (1919). These data can be presented in tabular form (table 12).

TABLE 12

Percentage of Genera in Which Woods Have
Helical Thickenings, by Floras

Region	Percent
Europe	60
Japan	40.7
U.S.A.	38.5
Taiwan	22.3
Java	15
Philippines	4.7

The data of table 12 show helical thickenings more frequent in higher latitudes. This tendency was also reported by Baas (1973) within the genus *Ilex*, and he cited similar data in other genera: *Celtis* (Schweitzer, 1971), *Gleditsia* and *Elaeocarpus* (Kanehira, 1921a, 1921b, 1924), *Euonymus* (Janssonius, 1906–1936), *Symplocos* and *Koelreuteria* (Baas,

1973). This is true in Asteraceae also: of species studied in this family, I found that 61 percent of temperate species— versus 46 percent of tropical species—have helical thickenings (Carlquist, 1966a). Also, the percentage in desert Asteraceae (68 percent) is much higher than in mesic Asteraceae (49 percent). Webber (1936) found helical thickenings exceptionally abundant in the desert shrubs she studied. Relatively primitive vessels, such as those of *Michelia* (plate 8-F), *Hydrangea* (Carlquist, 1961a) or *Ilex* (Baas, 1973), as well as highly specialized vessels (Asteraceae), can have helical thickenings.

The clear correlation here with latitude requires a physiological explanation. The most easily envisaged possibility is that helical thickenings offer resistance to collapse where higher negative pressures occur; this would accord with the presence of the thickenings in so many desert woods. However, a species such as *Michelia fuscata* hardly seems xeromorphic in any way. At least some woods with helical thickening have unusual flexibility and shear-resistance (*Ilex*). Nevertheless, this phenomenon remains one that will require further work, perhaps of an experimental nature, before any hypothesis can be supported.

PITTING ON IMPERFORATE ELEMENTS

Increasing division of labor between conducting and mechanical systems in dicotyledonous woods would be expected to feature loss of pit borders on imperforate elements. The figures of Metcalfe and Chalk (1950, p. xlv) support this trend in phylesis of dicotyledons at large. However, exceptions may be noted. In *Ilex*, temperate species have more numerous pits and wider borders on pits on tracheids than do the tropical species (Baas, 1973, p. 217). This situation may be related to the balance between the conductive capacity and mechanical capability of tracheids. The temperate species might possess greater conductive ability by virtue of imperforate-element pitting—an understandable correlation if conductive rates tend to fluctuate more in temperate zones, and if tracheids are still operable as a subsidiary conductive system in *Ilex*. Species of the temperate zone also tend to be smaller in plant size, in which case less stress would be placed on mechanically strong tracheids than in trees of the tropics.

A similar case is that of Icacinaceae. In vining species, fiber-tracheids and tracheids with numerous large bordered pits characterize scandent and lianoid species (Bailey and Howard, 1941). These authors claimed that these imperforate elements were primitive, and have lagged behind the imperforate elements of other Icacinaceae in evolutionary specialization. Even if this were true, one would still have to explain the cause. Vining dicotyledons at large typically show stem structure that stresses high conductivity (e.g., numerous, wide vessels) and poor mechanical strength (few imperforate elements per unit transection compared to vessels). If these principles are applied to vining Icacinaceae, one would expect imperforate elements to be precisely as they are: with numerous pits with wide borders for improved conductivity at the cost of loss of mechanical strength.

ROLE OF PARENCHYMA

Functions of parenchyma in xylem have been understood only tardily, and even today we need clarification. The functions of xylem parenchyma are probably more complex and subtle than those of tracheary elements. This dictates that a functional view is more difficult to achieve than with tracheary elements. The major contribution in this respect is Braun's (1970) book which, however, is directed primarily toward the development of structure-type schemes and only secondarily toward the exploration of physiological factors.

Axial Parenchyma

By means of staining reactions, Braun (1970) and Braun and Wolkinger (1970) have demonstrated that axial parenchyma cells are sites of photosynthate mobilization and deposition. Although this is most obvious in some temperate trees, it has been demonstrated for some tropical woods as well.

One must remember that nucleated fibers, discussed further below, are a substitute for axial parenchyma, and can have physiological equivalence to axial parenchyma. For example, Braun (1970) figures starch grains in "fibers" (tracheids?) of *Aucuba japonica, Berberis vulgaris,* and *Vitis vinifera.* Likewise, tracheids of *Gnetum* and *Ephedra* show phosphatase-test reactions and starch, indications of photosynthate storage and mobilization. Starch storage in ground tissue of monocotyledon stems such as the sago palm *Metroxylon* provides parallels.

Temperate trees have seasonality in growth that has an obvious correlate in photosynthate storage and mobilization. However, many tropical trees have sudden flushes of growth or flowering that require mobilization of photosynthates. These cycles may be related to the accumulation of photosynthates.

The cohesion-tension theory of water ascent in plants is noted by Braun (1970, p. 68) as a correlate of axial parenchyma. He cites this under the "*Albizzia* type" of organization, in which vessels are surrounded by broad sheaths of parenchyma. Braun feels that the parenchyma sheath must be especially large if water tension is particularly high in a plant. He believes that an incomplete parenchyma sheath around vessels, as in the *Aucoumea* type, would suffice to prevent occurrence of air embolisms.

Because, as cited earlier for conifer tracheids, air bubbles cannot pass through pit membranes, the value of a thick parenchyma sheath or an incomplete sheath with tracheids or libriform fibers facing a vessel is dubious in embolism-prevention, and such a potential function remains to be demonstrated. Complete and wide parenchyma cylinders around vessels characterize a minority of dichotyledons, and are not characteristic of dicotyledons known to experience high tensions in water columns, according to the data of Scholander, Hammel, et al. (1965). *Larrea divaricata,* with exceptionally

high measured tensions, has diffuse parenchyma (plate 15-A).
Encelia farinosa has incomplete sheaths of parenchyma around
vessels or vessel groups. This condition is also seen in *Artemisia
arbuscula* (plate 15-B), which probably experiences high tensions because of its habitat.

The universal occurrence of axial parenchyma or a substitute in dicotyledons seems related to conductive function or
closely allied metabolic activities. Crystals, tannins, resin-like
compounds, etc., accumulate in axial parenchyma. Läuchli
(1972) states, "lateral movement of ions from the xylem vessels into the surrounding tissues has been known to occur in
stems ever since its discovery by Stout and Hoagland." Or, as
Läuchli (1972) says with regard to various ions, "accumu-
lating experimental evidence indicates that there is a secretory
process implicated in the transport of ions from the symplasm
into the xylem vessels. This secretion is located in the xylem
parenchyma cells adjoining the vessels and is probably driven
across the plasmalemma by a carrier-mediated transport."

The evolutionary scheme of Kribs (1937) for evolution of
axial parenchyma, variously modified (Carlquist, 1961*a*; Braun,
1970) stresses various degrees of axial parenchyma cell aggregation with phyletic advance. These may represent progressively more effective modes of mobilizing and storing photosynthates, or translocating other compounds. One could say
that if grouping of parenchyma cells represents development
of tissue systems, rather than isolated cells, efficient translocation is achieved by any of various groupings. In this regard,
one might hypothesize that if the various trends of vessel evolution represent more effective, and therefore more rapid,
modes of sap conduction, these are paralleled evolutionarily
by progressively more efficient parenchyma conducting systems.

Substitutes for axial parenchyma.—Kribs (1937) hypothesized
that absence of parenchyma was primitive in dicotyledons,

but that axial parenchyma could also be lost secondarily. In fact, the vesselless dicotyledons have diffuse parenchyma, as do most families of dicotyledons with primitive vessel-bearing wood. Most of the families in which axial parenchyma is absent or very sparse are, in fact, rich in specialized characteristics in wood and other portions of the plant. Only a few are relatively primitive in wood anatomy: Violaceae is the only family that clearly falls into this category.

If we examine the list of families in which axial parenchyma is absent or very scarce, one finds that most of them have living imperforate elements (libriform fibers, etc.). Such families include Araliaceae, Berberidaceae, Cistaceae, Connaraceae, Euphorbiaceae (in part), Flacourtiaceae, Gesneriaceae, Lardizabalaceae, Myrsinaceae, Octoknemataceae, Polygonaceae, Rosaceae (*Spiraea*), Solanacaeae, Scrophulariaceae, Theophrastaceae, Verbenaceae, Violaceae, and Vitaceae. Nucleated septate fibers can be seen in such a wood as that of *Leonia glycicarpa* of the Violaceae (plate 12-A,B). Wolkinger (1969, 1970, 1971) has surveyed some woods with nucleated fibers and his results accord with the features observed in axial parenchyma. "Living" fibers are not only a substitute for axial parenchyma, they occur precisely where axial parenchyma is absent. This shift to nucleated fibers has occurred independently in various phylads of dicotyledons.

Nucleated fibers, if they represent an alternative to axial parenchyma in storage and conduction, represent a very efficient innovation on account of the fact that massive groupings of cells are involved, much like a total sheet of parenchyma, although perhaps not so efficient per cell as non-lignified parenchyma. As an alternative to aggregations of axial parenchyma such as apotracheal or paratracheal, it seems a quite good alternative, combining the best features of both.

The families Punicaceae and Sonneratiaceae lack axial parenchyma, as do a few Onagraceae. These families, however, have interxylary phloem; if conduction of photosynthates within the xylem is a function of axial parenchyma, a more apt substitute than phloem (which also contains some parenchyma cells) can hardly be imagined.

Also, the families with parenchyma absent or scarce tend to have rays elongate, wide, or numerous, as in *Leonia* (plate 13-A,B), so that vertically elongate ray cells might well substitute for axial parenchyma and certainly form good contacts between ray parenchyma and nucleated fibers. In this connection, one may note that in primitive dicotyledons, such as *Amborella* (plate 6-B), *Drimys winteri* (plate 7-B), or *Illicium cambodianum* (plate 9-F), multiseriate rays have long wings of upright cells, and upright cells are present in uniseriate rays. These would form good contacts with diffuse parenchyma in these species.

Rays

As in axial parenchyma, starch and phosphatase presence in rays of dicotyledons (Braun, 1970, p. 142) indicates that rays perform a significant function in storage and radial translocation of photosynthates. This fact should be stressed, despite the fact that only a few rays have been studied in this respect.

Radial translocation of ions is also accomplished by rays. As Läuchli (1972) states, "the rays may also function in mediating transport of ions between the two channels [xylem and phloem] of longitudinal translocation." Läuchli also offers more specific examples: "calcium is also exported from the xylem vessels of the stem and may be accumulated irreversibly as Ca oxalate, or moved toward the periphery of the stem where it is deposited in the epidermis and in hypodermal tissues." Other ions also experience radial translocation according to Läuchli (1972): "in willow, some accumulation of such ions as K^+, Na^+, SO_4^{--}, and PO_4^{---} takes place at first in tissues between the vessels of the xylem and the sieve tubes of the phloem."

Presence of metabolic byproducts (crystals, resin-like compounds, tannins, etc.) in ray cells suggests a prominent excretory function, and few ray cells in dicotyledons are totally devoid of such substances.

Rays viewed conjunctively with axial parenchyma.—In connection with the translocation of both photosynthates and ions, ray types are of significance. Unfortunately, in neither Kribs's (1935) nor Braun's (1970) survey is there a contrapuntal view of ray and axial parenchyma types.

For example, diffuse parenchyma tends to be associated with Kribs's "Heterogeneous I" rays (fig. 1), as in *Illicium* (plate 9-E,F). In "Heterogeneous I" rays, upright cells are numerous, along the sides of multiseriate rays, in the wings of multiseriate rays, and in uniseriate rays. Axial parenchyma cells are long, corresponding to the length of fusiform cambial initials. Thus, even though axial parenchyma cells are diffuse, there are numerous points of contact between vessels and them, and between them and ray parenchyma cells. Also, tracheids have a subsidiary conductive function in a primitive vessel-bearing wood such as this, and thus direct contact between vessels and axial and ray parenchyma is not important.

In this regard, one may note that in conifers and vesselless dicotyledons, axial parenchyma is quite sparse, but tangential transfer of materials would be accomplished readily, by the numerous pits in radial walls of tracheids, into procumbent ray cells.

In primitive dicotyledons, upright cells are relatively abundant in rays; these form a broad base of connection with axial parenchyma and tracheary elements, as mentioned; if procumbent cells denoted efficient radial translocation, the primitive dicotyledons would, by the relative paucity of ray cells, be expected to have moderately slow radial translocation, just as diffuse parenchyma would represent relatively inefficient axial translocation. However, if so, the translocation capability of xylary parenchyma in a primitive wood would be matched by moderate conductive rates in tracheary elements.

An example at an opposite extreme would be represented by

Crataeva. Crataeva was cited by Kribs (1935) as an example
of his "Homogeneous II" type of rays (fig. 1). In "Homogeneous II," only broad multiseriate rays are present. All ray cells
are procumbent. *Crataeva* has aliform-confluent rays (Metcalfe and Chalk, 1950). These are broad sheaths of parenchyma surrounding vessels, with the sheaths tending to contact
each other tangentially. Thus, in *Crataeva*, vertical sheaths of
axial parenchyma intersect broad rays; both parenchyma systems are relatively efficient for longitudinal and radial translocation, respectively. These are matched by efficient vessel-
element conductive capability; *Crataeva* has wide vessels with
simple perforation plates. In addition, imperforate elements
in *Crataeva* are all libriform fibers, and thus non-conductive,
so that effective contact of vessels only with parenchyma occurs where translocation from vessels to parenchyma is concerned, with no tracheids to serve as intermediaries. However,
the broad sheets of axial and ray parenchyma, which intersect
each other, assure efficient (and rapid) translocation of a
compound from vessels via axial parenchyma to the rays, or
from xylem to phloem or from phloem to xylem.

These two isolated examples demonstrate two points in an
infinitely complex picture. Ideally, one would wish to achieve
a linking of the increasing division of labor in the tracheary
elements with the increasing massing of axial and ray parenchyma into interfaces (both with each other and with the
tracheary-element system), with all systems evolving progressively greater conductive and translocation efficiency. This
transition in tracheary element and parenchyma systems
should be viewed three-dimensionally, rather than as transverse, tangential, and radial sections, and it should be viewed
in terms of phylogenetic specialization within orders, families
and genera of woody dicotyledons, and their evolution in ecological capabilities. This prospect obviously requires a preposterously complex synthesis.

To illustrate further the complexities, one would have to
take into account vessel diameter, abundance and distribution.

In Asteraceae, for example, rays are mostly narrow multi-seriate; axial parenchyma occurs as incomplete sheaths around vessels. This would make for inadequate contact between the vessels, the axial parenchyma, and the ray parenchyma except that vessels tend to be narrow, numerous, and grouped, so that contacts to rays are assured even with relatively small amounts of axial parenchyma.

Mechanical capabilities of rays.—In many species of dicotyledons, ray cells bear thick, lignified secondary walls. Because of the relative volume that rays occupy in woods, the potential strengthening effect of ray cells with such walls is very considerable. This should be taken into account when considering the mechanical strength of a wood, for it may materially add to strength furnished by the axial xylem. Axial parenchyma may bear thick lignified walls also, but probably adds less to mechanical strength than do rays.

Bordered pits on ray cells were claimed by Kribs (1935) to be primitive in dicotyledons. However, bordered pits on ray cells may be found in such diverse taxa as *Drimys winteri* (plate 7-D), with primitive wood, and *Metrosideros* (Sastrapadja and Lamoureux, 1969), which has a specialized wood. Bordered pits in ray cells could be explained as a mechanism to maximize mechanical strength of ray cells while retaining translocation capability among ray cells. This formula could be applicable in woods of various types.

A potential function of rays contradictory to the above is the possible compressibility of rays during expansion and contraction of the stem. MacDougal (1921) demonstrated these diurnal contractions, based upon the increased tension in xylem during hours of high transpiration. Relatively thin-walled ray cells would accommodate these diurnal contractions and expansions most readily. A related but different phenomenon is the occurrence of large rays (often little modified from pith rays) in wood of such succulents as Cactaceae and Crassulaceae. These large zones of thin-walled cells provide sites for shrinkage and expansion during the rather drastic changes in volume during wet and dry seasons. The various roles of ray cells, pith, cortex, and axial parenchyma in dicotyledonous stem succulents could easily be elaborated at this point. Incidentally, the structure of monocotyledonous stems is pre-adapted to expansions and contractions of ground tissue without harm to vascular bundles (often enclosed in bundle sheaths in any case); monocotyledonous stem succulents are ideally adapted to fluctuations in stem and leaf volume.

Flexibility of rays is a significant factor in construction of dicotyledonous lianas and vines. If normal cambial activity is present, flexibility is achieved by formation of tall, wide rays which are little-modified extensions of pith rays.

CONDUCTIVE CHARACTERISTICS OF WOODS
OF VARIOUS ECOLOGICAL TYPES AND GROWTH FORMS

In figure 14-B, we can see a straight-line relationship between the vessel diameter and the number of vessels per square millimeter, based on an assortment of mesic species (dots), some with simple perforation plates, but most with scalariform perforation plates. Although vessel diameter varies widely in these species, its relationship to the number of vessels per sq. mm. is always very close to inverse, so that in essence, the conductive area is constant in these species. The alternatives of numerous, narrow vessels versus few, wide vessels are not without significance. For example, wide vessels might be of selective value for plants with large leaves (e.g., Myristicaceae, which are dots at the wide-vessel end of the curve in fig. 14-B) or other circumstances which might show a requirement for rapid conduction of large volumes of water per unit time. Obviously, vines ("V" in fig. 14-B) represent another such instance: vessel diameter is much wider in lianoid than non-lianoid Passifloraceae, for example (Ayensu and Stern, 1964). Such correlations form the basis for the discussion below.

Mesic Woody Species

If we take an assemblage of mesic woody dicotyledons as an arbitrary "standard," we can compare other growth forms against them, as has been done in figure 14-B and table 13. Using species in these categories from my collection of wood-section slides, I have calculated for each species the average vessel-element length, average vessel diameter, and average number of vessels per sq. mm. I then calculated a figure for the conductive area of vessels in a square mm. of transection, using average vessel diameter to find the transectional area of the average vessel, then multiplying that figure by the number of vessels per sq. mm. The resulting figure is not precise, because measurement of vessel diameter as typically done by plant anatomists includes wall in addition to lumen and is based on the widest diameter of each vessel measured. If not precise, this figure is calculated the same way for each species, and is comparable within limits. The figures obtained for each feature of the species in each plant type category were then averaged, so that the figures in table 13 represent "averages of averages." The symbols in figure 14-B represent data for species individually.

Vines and Lianas

Vines and lianas appear to show the least deviation from the pattern of woody mesic species in conductive area in table 13, even though they tend to form a distinctive grouping on figure 14-B. The position of the vining species in figure 14-B suggests that they might have lower conductive area than the woody species; in fact, as table 13 shows, conductive area per sq. mm. is greater, as one would expect. The method of measurement (including wall, rather than just lumen) makes for a slight bias toward higher conductive area with smaller vessel diameter.

The exceptionally wide vessels of vines have minimal friction

TABLE 13

Characteristics of Conductive Tissue in Stems of Dicotyledons
Grouped According to Habit and Habitat

Plant Type	Vessel Diameter (Average)	Vessels per Sq. mm. (Average)	Total Vessel Area per Sq. mm. (Average)	Vessel-Element Length (Average)
Primitive woody				
mesic species	108 μ	47.2	.242 sq. mm.	1385 μ
Rosette trees	79 μ	30.6	.140	412 μ
Vines & Lianas	157 μ	19.1	.359	334 μ
Annuals	61 μ	162	.324	186 μ
Desert shrubs	29 μ	353	.178	218 μ
Stem succulents	72 μ	64.2	.087	259 μ
Woody, but with				
successive cambia	68 μ	16.6	.058	146 μ

to water movement and are therefore far more efficient than conductive area alone would suggest. Also, vessels tend to increase in diameter in vines as the stem increases in diameter (fig. 16). The selective value of a wide vessel is positive only where high negative tensions in water columns do not exist. The combination of a large volume of water conducted per unit time with moderate negative pressures per unit time is relatively rare, and explains why most dicotyledons have not evolved progressively wider vessels. The vessel-element length of vining species (table 13) seems long compared to other groups, but the length is only average in dicotyledons as a whole (fig. 13-A). If one notes that the length is less than twice the width for an average vessel element of a liana or vine, the length seems almost minimal: the ratio is less than for other plant types. Having the lowest ratio is logical for vining species, because short length probably does enhance resistance to negative pressures.

Presence of numerous, wide vessels with relatively little me-

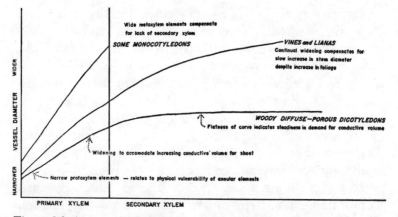

Figure 16. Age-on-vessel diameter curves for three categories of angiosperms. The curve for "some monocotyledons" is based on *Washingtonia filifera* (Arecaceae). The curve for "vines and lianas" is based on *Doxantha unguis-cati* (Sapindaceae). The curve for "woody diffuse-porous dicotyledons is based on *Scaevola taccada* (Goodeniaceae).

chanical tissue characterizes scandent species but would be an unsatisfactory xylary formula for a tree. In a vine or liana, mechanical strength has no positive value where self-support is concerned, whereas this is a major value of mechanical strength in a tree or shrub. Vines do have mechanical tissue of great tensile strength—often as extraxylary fibers—and fibers are often very long. Mechanical tissue in a vine may serve to some extent to prevent rupture of vessels under torsion.

Stem Succulents

Stem succulents do seem to fall appreciably below the mesic woody species in conductive area per unit transection (table 13). This would have been dramatized had a normal instead of a log-log scale been used in figure 14-B, but it is still quite evident in that graph. Stem succulents have an average vessel diameter more than twice that of the desert shrubs, but a low number of vessels per unit area (although the average is deceptive, for the stem succulents are scattered rather widely on fig. 14-B). However, the average figure does tend to illustrate

the expected stereotype of a stem succulent in that it indicates
relatively few vessels, abundant parenchyma (cortical and pith
tissues are not, of course, included in the data of table 13).
Although vessels are narrower in stem succulents than in the
average dicotyledon (see fig. 13-B), the difference is very small.
This is exactly what we might expect, because succulents have
low transpiration rates yet moderate negative pressures, as
cited earlier in this chapter. A succulent is a system for rapid
absorption of water on a seasonal basis, but with a slow but
steady rate of transpiration. Roots can die back to cork-pro-
tected tissues during the dry season. The low conductive area
per unit transection seems directly related to low transpiration.

Desert Shrubs

The vessels of desert shrubs are very narrow (table 13), as
one would have expected from the fact that narrow vessels
resist tension in water columns better than wider ones, and
from the fact that desert shrubs have the highest negative
pressures of any type of plant (Scholander, Hammel, et al.,
1965). The conductive area per unit transection is lower than
for mesic dicotyledons (fig. 14-B). This would be expected
on the basis that transpiration from a desert shrub is lower
than that of a mesic shrub or tree. There are, however, exceptions to this, such as *Artemisia tridentata* (vessel diameter,
28 μ; vessels per sq. mm., 747; area of vessels per sq. mm.,
.505 sq. mm.). The most extreme example of the desert shrub
type of wood construction I know of is not from a desert, but
an alpine region, *Loricaria thuyoides* (plate 15-C,D), although
it is less seasonal on account of its equatorial latitude. It grows
in the superpáramo of Colombia. Other equatorial alpine
Asteraceae, such as *Stoebe kilimandscharica*, show tendencies
in this direction (Carlquist, 1961b). Strong seasonality is indicated in woods of most desert shrubs, like the *Artemisia* (illustrated in plate 15-B). Data on wall thickness of vessels
would be interesting; desert shrubs tend to have thick-walled
vessels (plate 15-A,B). However, because of the interrelation-

ship between diameter and vessel wall thickness in strength characteristics, an investigation would have to take the form of developing a proportion, and noting which plants have vessels that are notably thick walled, and which are notably thin walled.

The microphyllous or aphyllous condition of desert shrubs dictates that smaller volumes per unit time may be conducted in a stem, and that the conductive rate would be slow. Narrow vessels are not disadvantageous if conductive rate is slow, because the friction they offer does not become appreciable when conduction is slow.

Annuals

Numerous narrow vessels or vascular tracheids can be seen in the last-formed wood of an annual, or an annual stem of a perennial, as in *Aster spinosus* (fig. 17-A). Thus a stem of an annual may begin with a wood structure like that of a forest tree but end with wood like that of a desert shrub. Therefore, averages for quantitative xylem characteristics in annuals may not be entirely meaningful. However, the fact that annuals have, on the average, a greater conductive tissue than mesic woody species (table 13) is understandable. Annuals grow during the mesic portion of a year only, with a few exceptions, so transpiration from an actively growing, broad-leaved shoot might be expected to be great, with a concomitantly greater conductive area in transection of xylem. The shortness of vessel elements in annuals takes on significance if, at the end of a growing season, the vessel elements are also narrow, and thus equipped to conduct under high tension, a capacity that might be related to the capability of an annual to complete its life cycle despite progressive drying of soil.

Rosette Trees

The rosette trees in table 13 have a vessel diameter typical of mesic woody dicotyledons. Rosette trees transpire at a steady rate, a fact related to their constant quantity of leaves and their lack of marked increases or decreases in growth rate. Conductive area for rosette trees is lower than that for woody dicotyledons, but not as low as that for stem succulents (table 13). The rosette trees surveyed (only unbranched trees were included) are rather like stem succulents, so perhaps one should not be surprised that their xylem is rather like that of a succulent. Such a tree as *Carica* could be classified as a rosette tree or a stem succulent on the basis of its xylem anatomy, and gross morphology of the plant would also make it intermediate between these two categories.

Parasites

Woody dicotyledonous parasites have, as mentioned, short, thick-walled vessels, and offer close parallels to woods of species in xeric regions. However, angiosperms parasitic on mycorrhizae or angiosperms parasitic on roots of other angiosperms offer other patterns.

In mycoparasitic dicotyledons, one might expect a lack of vessels, if the patterns seen in the monocotyledons *Petrosavia* and Triuridaceae were applicable. However, mycoparasitic dicotyledons are mostly different in habit and habitat from those monocotyledons. The most conspicuous group, the subfamily Monotropoideae of Ericaceae, does have vessels (Metcalfe and Chalk, 1950; Bierhorst and Zamora, 1965). *Petrosavia* and Triuridaceae are small plants with a slow growth rate, whereas Monotropoideae are larger, succulent, and develop massive inflorescences and fruits within a relatively short period of time. Thus, rapid conductive rates during growth and therefore presence of vessels (and also specialized sieve tube elements) would be expected.

Similar considerations apply to the root parasites Balanophoraceae, Orobanchaceae, and Hydnoraceae, all of which have vessels (Metcalfe and Chalk, 1950). These all produce flowers and fruits rapidly, and rapid conduction in xylem and phloem would be expected.

Rafflesiaceae are root parasites that have tracheids, but no vessel elements in xylem so far as is known at present (Metcalfe and Chalk, 1950). Rafflesiaceae are tropical and subtropical endoparasites in which flowers develop relatively slowly. In the more temperate regions, flowers are small (*Apodanthes, Pilostyles*). Probably slow development is related to slow conductive rates, and therefore vessels are not advantageous. One may hypothesize that Rafflesiaceae are a parasitic offshoot of a line in which only tracheids are present in primary xylem. Such primary xylems do occur in a number of dicotyledonous families (Bierhorst and Zamora, 1965).

Anomalous Secondary Growth

Plants with successive cambia have a low figure for conductive area per unit transection (table 13) because parenchyma between the vascular bands was included. Had only the vascular bands been calculated, the figure would approach that of normal woody plants. The very short element length in these "anomalous" woods is unexpected unless one remembers the site of origin of these cambia. Whereas the cambium in a normal woody plant begins with vertically elongate undifferentiated procambium, the successive cambia originate from cortical parenchyma cells, which are very short in comparison. The products of successive cambia are thus very short, and cambial initials change very little in length during the duration of activity of each cambium. Such short initials would limit the length of imperforate elements except that in these woods, imperforate elements can have great intrusiveness—they may be five (*Nototrichium*) or six (*Charpentiera*) times the length of vessel elements.

The formation of successive cambia is a form of secondary activity that has been attained more than once in dicotyledons. Whatever the phylogenetic routes by which it has been achieved, some growth forms are more predisposed to it than others. Groups with successive cambia tend to be small to intermediate sized trees at best (*Charpentiera, Pisonia, Nototrichium, Phytolacca dioica*). Where arborescent, a form of mechanical strength can be found, as with the unusually long, thick-walled fibers of *Nototrichium* and *Charpentiera*. *Phytolacca dioica* has an unusual mechanism: the "grain" of each band of vascular tissue is oriented obliquely—about 30–45° from the vertical—with each band alternating in direction (clockwise versus counterclockwise) of twist from the one preceding it (David Wheat, personal communication).

Shrubs occur in some groups with successive cambia (*Russchia, Atriplex*), but lianas contain a higher proportion of species with successive cambia than do other growth forms. Lianas with normal secondary growth feature a minimal accumulation of secondary xylem and a high proportion of ray parenchyma in stems. These features are not at all incompatible with the successive cambial mode of growth, in which parenchyma between and among bands functions like ray tissue. Vascular bands, in the successive-cambia species, have the advantage that each is encircled by parenchyma, so that the flexibility of "cable" construction is achieved. Stems of lianas, even those with normal cambial activity, are often asymmetrical in transection. This is related partly to the lack of self-supporting characteristics, but also to their addition of vascular tissue in relation to variously leaning on, twining up, or lying upon trees. The successive cambial mode is exceptionally well suited to asymmetrical development.

Species with less parenchyma between the bands of vascular tissue can be said to mimic more closely the xylem produced by normal cambia. To the extent this is true, woodier forms tend to occur (e.g., *Cocculus laurifolius*). One can regard the opposite—that is, wide parenchyma bands between cambia—as predisposing to succulence and storage functions (roots of *Abronia, Beta,* and *Boerhavia*). Highly parenchymatous secondary xylem occurs in a number of trees with normal cambial activity (*Erythrina, Carica*), so successive cambia merely reproduce from a different source the parenchyma that would be formed from a normal cambium.

Primary Xylem of Dicotyledons

Age-on-length curves, such as those of figures 15 and 17, are now familiar to plant anatomists. The significance of the secondary xylem portion of these growth curves has been discussed earlier in this chapter. However, the nature of changes in the primary xylem needs further discussion. What is the reason for longer protoxylem than metaxylem elements (fig. 15)? The only known exception to that trend, in cycads, has been discussed in chapter 6.

The great length of protoxylem elements at the beginning of the age-on-length curves is explicable as the result of elongation in procambial initials; there is, in addition, stretching that occurs after maturation of the protoxylem elements. Metaxylem elements would be just as long, in a given stem, as protoxylem elements, if elongation in procambial initials alone governed the length of metaxylem elements. The elongation that governs the length of the protoxylem elements both before and after their maturation would be operative. However, metaxylem elements are at least a little shorter than protoxylem elements, and in most woody plants markedly shorter. This is caused by transverse divisions in procambial cells destined to become metaxylem. Intrusiveness of tips of xylem cells may occur during maturation (Bierhorst and Zamora, 1965), but to a limited extent.

In functional terms, the value of the increased diameter in metaxylem elements compared to those of protoxylem (fig. 16) is an accommodation of conduction of larger volumes of water to a shoot system as it leafs out and begins transpiring rapidly. What functional reasons for short metaxylem elements can be cited, particularly when tracheary element length begins to rise again at the beginning of secondary growth in conifers and dicotyledons? There must be a positive value for shorter elements, just as one must hypothesize this to explain continued phyletic shortening of vessel elements in woods of dicotyledons (e.g., table 10). The explanation offered previously appears valid in the case of metaxylem: the constriction, forming a non-pitted rim, at each end of a metaxylem vessel element, adds considerably to resistance to collapse. This is even more true in the case of a tracheid, where an entire end wall is interposed as a strengthening mechanism. Retarded introduction of vessels into primary xylem in dicotyledons (Bailey, 1944b) would accord with value of a stronger end wall in metaxylem. If end walls of metaxylem elements, whether tracheids or vessel elements, tend to provide strengthening, the more such walls per unit length of vessel, the greater the resistance to collapse.

If wider vessels are mechanically stronger when shorter, we would expect shorter vessels in the earlywood of ring-porous woods. This does occur (Swamy, Parameswaran, and Govindarajalu, 1960). Those authors attribute this to the occurrence of pseudotransverse divisions in the cambium at the beginning of the ring; it might better be attributed to the greater intrusiveness of latewood vessels (which are more nearly fusiform in shape) during maturation. This would also explain why curves given by those authors for perforate and imperforate elements in a wood are not always parallel.

If wider vessels are mechanically stronger when shorter, one ought to expect that ring-porous woods at large would have short vessel elements, for the wide vessels at the beginning of growth rings would dictate the maximal length congruent with strength required by wide vessels. Interestingly, ring-porous woods do have shorter vessel elements. For example, I selected three tribes of Asteraceae and from my published data on them, averaged all species according to whether they were ring porous or diffuse porous. In the diffuse-porous species, vessel elements averaged 303 μ in length; in the ring-porous species, 236 μ. Interestingly, extreme diameter might be more significant than average vessel diameter in control of this situation, for the same species, with respect to vessel diameter, averaged 89 μ in diffuse-porous species, 71 μ in ring-porous species. However, the narrower diameter in ring-porous species could, instead, be influenced by the fact that there are more numerous latewood vessels than earlywood vessels. Similar data for length and diameter of vessel elements in diffuse versus ring-porous woods could be provided for other families; for example, Anacardiaceae (David Young, personal communication).

In those dicotyledons in which division of labor is maximal, one might expect that the length of metaxylem vessels would continue with little change into secondary xylem. This is, in fact, true, as in *Swietenia*, for which a curve is given by Bailey and Tupper (1918). Graphs are given for tracheid length in various conifers by Bailey and Tupper (1918). In all of these, there is a sharp upswing from metaxylem tracheid length into secondary xylem with successive years. Here, mechanical and conductive functions are both maximized by greater length, so this is understandable. If, in a specialized dicotyledon, the fusiform cambial initial length increased at the onset of cambial activity and (through lack of intrusiveness) the imperforate elements were the same length as the vessel elements, either the vessel elements would be weak by having greater length, or mechanical strength would be low by virtue of shortness of the imperforate elements. Because of division of labor, the lengths of vessel elements and imperforate elements are relatively independent: vessel-element length becomes shorter with phylesis, libriform fibers gain greater intrusiveness (even though they do not become as long as imperforate elements in primitive dicotyledons). Among specialized dicotyledons, there is, according to the data of Bailey and Tupper (1918), one species in which imperforate-element length is approximately the same as vessel-element length: *Leitneria floridana*. *Leitneria floridana* is a short, coppicing shrub of perpetually wet streambanks. With respect to plant size, long imperforate elements have less selective value; with respect to tensions within vessel elements, they are likely to be low because of an unlimited supply of water. Either factor would permit the equivalence in length of perforate and imperforate elements.

Morphology of Primary Xylem Elements

Bierhorst (1960) and Bierhorst and Zamora (1965) have presented us with a survey of primary xylem patterns that demonstrates great variety in structural modes. The particular types of tracheary elements they illustrate—particularly in metaxylem—are unusual and invite explanation. These types are too numerous to permit point-by-point correlation with adaptational significance here. More importantly, that cannot be done, except in a speculative way, unless one knows well the anatomy of the plant in which the tracheary elements are located, as well as its growth form and ecology. For example, do metaxylem elements with wide gaping pits occur in plants with lowered mechanical strength in xylem, with compensation by means of extraxylary fibers? Such questions will require detailed studies before answers can be formulated, or the trends proposed by Bierhorst and Zamora (1965) can be validated or interpreted in adaptive terms. However, two unusual primary xylem types are considered below to indicate possibilities.

Nepenthes has odd helical elements in which cell tips have lignified caps (Bierhorst and Zamora, 1965). This seems related to the tendency in *Nepenthes* to attain great mechanical strength in primary stems. Xylary and extraxylary fibers in *Nepenthes* stems are very strong, and stems are very difficult to break by hand. *Drosera* is unusual in having vessel-elements in which a circular perforation plate is present, but the perforation occupies only a small portion of the end wall (Bierhorst and Zamora, 1965). One could hypothesize that *Drosera* is derived from a phylad in which perforation plates in primary xylem are simple. Conductive rates in *Drosera*, an acaulescent (in the species studied by those authors) bog plant, are probably low, so a small perforation plate might suffice. Such detailed speculation is obviously premature. Moreover, variabilities in structural details of primary xylem are great, and not all of these variations may have adaptive significance.

PAEDOMORPHOSIS

The main facts regarding morphological and ontogenetic aspects of paedomorphosis were developed in my 1962 paper, and a number of examples have been cited (Cumbie, 1963, 1967a, 1967b; Anderson, 1972; Gibson, 1973). That this phenomenon is related to herbaceous plants, plants secondarily woody from an herbaceous ancestry, stem succulents, rosette trees, and allied specialized situations and is not generally applicable to woody dicotyledons at large is clear. More than one outcome exemplifies the phenomenon, or better, phenomena, of paedomorphosis (fig. 17). These outcomes represent various alternatives in herbaceous structure.

The "most herbaceous" of these outcomes can be considered an annual, a short-lived perennial, or an annual stem of a perennial plant, as in the case of Aster spinosus (fig. 17-A). In Aster spinosus there is a drop in tracheary-element length from protoxylem to metaxylem, but then a continual decrease in length. One explanation consistent with the facts is that there is a decrease in tracheary-element length with release in mechanical strength. If shorter elements were of any value on account of a decrease in vessel-element length and a concomitant decrease in vessel diameter, we should see a consistent decrease in vessel diameter outward in the stem. However, there is in a transection no decrease in vessel diameter until the final millimeter or less of growth. That final decrease undoubtedly is related to increasing tension in xylem, as mentioned earlier in this chapter.

Release from mechanical strength is also shown in the curves for Talinum guadalupense and Macropiper excelsum (fig. 17), both of which can be considered stem succulents.

Macropiper has "less release," which accords with its taller stature. *Macropiper* also has libriform fibers in secondary xylem, but wide, tall rays not broken into segments by secondary growth. *Talinum guadalupense* shows even greater release from mechanical strength not only in the decrease in its length-on-age curve, but also in the complete lack of any fibers or other sclerenchyma in the secondary xylem. Both *Talinum guadalupense* and *Macropiper excelsum* have lateral wall pitting on vessels of a scalariform or scalariform-like type; this denotes a lack of mechanical strength. If trends such as these are true in these stem succulents, they ought to occur in the largest group of dicotyledonous succulents, Cactaceae. In fact, all of these trends can be found in cacti (Gibson, 1973).

Ontogenetic Aspects of Paedomorphosis

In formal ontogenetic terms, the procambium of *Talinum guadalupense*, *Macropiper excelsum*, and *Aster spinosus* begins with relatively short cells compared with procambium of typical woody dicotyledons. Then horizontal or somewhat diagonal divisions decreasing procambial and fusiform initial length take place. Thus these elements shorten as the stem

Figure 17. Age-on-length curves for woods exemplifying types of paedomorphosis with the addition, in A, of three species with a "normal woody plant" curve. In B, those three species have been omitted, and the types of paedomorphosis shown by the species in A have been categorized. Although tracheid lengths (*Pseudotsuga* and *Dioön*) and vessel-element lengths (others) have been used for this graph, lengths of libriform fibers would have been more indicative of degrees of mechanical strength in the dicotyledons. The graph is based on vessel elements because *Talinum guadalupense* lacks libriform fibers, because vessel elements can be measured much more accurately than libriform fibers in radial sections, and because vessel-element lengths are a reliable indicator of lengths of fusiform cambial initials. Lengths of libriform fibers can be assumed to parallel those for vessel elements. (Curves for *Pseudotsuga menziesii* and *Dioön spinulosum* modified from Bailey and Tupper, 1918; curve for *Idria columnaris* modified from Henrickson, 1968; curves for *Talinum guadalupense* from Carlquist, 1962; others are original.)

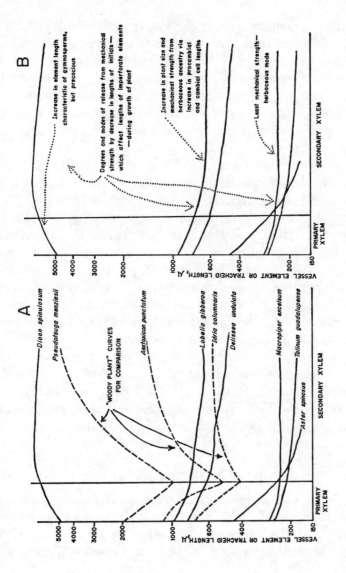

B

Increase in element length characteristic of gymnosperms, but precocious

Degrees and modes of release from mechanical strength by decrease in lengths of initials— which affect lengths of imperforate elements —during growth of plant

Increase in plant size and mechanical strength from herbaceous ancestry via increase in procambial and cambial cell lengths

Least mechanical strength— herbaceous mode

PRIMARY XYLEM   SECONDARY XYLEM

VESSEL ELEMENT OR TRACHEID LENGTH, μ

150 200 400 600 2000 3000 4000 5000

A

Dioon spinulosum

Pseudotsuga menziesii

"WOODY PLANT" CURVES FOR COMPARISON

Acrtoxicon punctatum

Lobelia gibberoa

Idria columnaris

Delissea undulata

Macropiper excelsum

Talinum guadalupense

Aster spinosus

PRIMARY XYLEM   SECONDARY XYLEM

VESSEL ELEMENT OR TRACHEID LENGTH, μ

150 200 400 600 1000 2000 3000 4000 5000

grows in diameter. These divisions occur less frequently in the species cited than they would in typical woody plants, hence the curves of figure 17-A, which show a descent less steep than for woody species. These divisions, however, continue into secondary xylem, without the occurrence of an increase in the length of fusiform cambial initials (not to be confused with *derivatives* of fusiform cambial initials) by means of gradual intrusive growth.

Lobelia gibberoa and *Delissea undulata* (fig. 17-A) also show paedomorphosis curves. However, they begin with longer tracheary elements than do the species cited above. Paedomorphosis is not responsible for greater length of the first-formed elements. Rather, one may assume that the selective value of longer libriform fibers in these species at the outset of growth is responsible. However, pseudotransverse divisions do take place, although at a slow rate. Relatively few transverse divisions take place in ray initials also, since whatever governs division in one part of the cambium is evidently effective in another. Consequently, these two lobelioids have a high percentage of erect ray cells. A high proportion of erect ray cells, with few or no procumbent cells, is typical of woods that show paedomorphosis (Carlquist, 1962, 1966a, 1969a, 1969b; Gibson, 1973). This ray type is not included in the system of Kribs (1935), presumably because he studied no species in which it occurs.

Functional Significance of Paedomorphosis

Why should rosette trees such as *Lobelia gibberoa* and *Delissea undulata* be examples of "release" of mechanical strength? First, they are unbranched, so that the addition of only a moderate amount of secondary xylem suffices to maintain mechanical strength. The amount of foliage increases little, so that only the support of the increasing length of stem, and resistance to wind thrust on that much, are factors influencing the amount and strength of xylem. In fact, rosette trees such as lobelioids are all of moderate height, rarely exceeding 20–25 feet, and mostly much shorter. Libriform fibers of progressively shorter length, and with rather thin walls, obviously suffice. The lack of marked mechanical strength in the lobelioids is also confirmed by vessel wall pitting: pits are large, often elongate, sometimes simulating a scalariform pattern.

If progressive shortening of fusiform cambial initials represented a physiological value by virtue of shorter vessel elements, more resistant to collapse, one would expect to see a decrease in vessel diameter, as well as mechanically strong vessel walls. However, these do not occur in the lobelioids, or the other examples cited earlier.

Patterns suggesting a decrease in mechanical strength of woods in species other than those that show paedomorphosis can be cited. In *Fraxinus excelsior,* a decrease in libriform fiber length after the first 30 years was observed (Bosshard, 1951). Terminal decrease in tracheid length in old trees was reported in *Pinus densiflora* by Hata (1949) and *Sequoia sempervirens* (fig. 7) by Bailey and Faull (1934). That such a decrease in tracheary-element length in these woody species takes place so tardily suggests that there is a protracted value for mechanically strong elements in xylem as compared to wood of herbaceous plants.

There may seem a certain irony in the fact that in the lobelioids cited, phyletic entry into mesic sites and concomitantly greater arborescence have resulted in an increase in tracheary-element length compared with herbaceous ancestors; yet in the ontogeny of a stem of a tree *Lobelia,* there is decrease in mechanical strength. As noted earlier in this chapter, certain circumstances do permit herbaceous phylads to become arborescent. Some of the products of this secondary arborescence are growth forms such as cacti and rosette trees in which the mechanical strength of xylem is not of marked selective value. In these cases, paedomorphosis can be expected. If radiation of an herbaceous group resulted in typical trees, few or no features associated with paedomorphosis would be expected.

Another seeming paradox is that one form of paedomorphosis carries with it an increase in mechanical strength: raylessness. As I have stated (1970c), raylessness can occur when an herbaceous phylad (1) increases in woodiness; (2) has very short fusiform cambial initials; (3) has, as do many stems at the beginning of primary growth, erect ray cells. Under these circumstances, if pseudotransverse divisions do not occur, the longitudinal length of ray cells is approximately equal to the length of imperforate cells in fascicular portions of the xylem. If both ray cells and imperforate fascicular cells remain thin walled, a stem succulent like that of Crassula argentea (plate 13-C,D) can result. If both ray cells and imperforate fascicular cells mature into fibers, a rayless wood like that of Aeonium arboreum results. The rayless pattern features absence of transverse divisions in areas of the cambium corresponding to ray initials—or else tardy appearance of such divisions, in which case rays occur in the secondary xylem. The absence of transverse or pseudotransverse divisions in the cambium of rayless woods can be likened to the slow pace of pseudotransverse divisions in woods that show paedomorphosis. The nearly flat curve illustrated for Macropiper excelsum in figure 17-A is indicative of the virtual absence of pseudotransverse divisions. Macropiper is not rayless, because it does not satisfy conditions (2) and (3) above.

𝒞𝑒 12.

Sieve Elements;
Extraxylary
Mechanical Tissue

TRANSLOCATION IN PHLOEM
AND SIEVE-ELEMENT MORPHOLOGY

The functional correlates of xylem evolution invite comparisons with phloem. If one analyzes the data of Cheadle and Whitford (1941) on the phloem of monocotyledons, one finds that simple sieve plates occur in organs where rapid translocation occurs: leaves and storage organs. In rhizomes, etc., mobilization of photosynthates may be rapid during the beginning of growth during a brief season; input of photosynthates for storage would also tend to be rapid. In organs of monocotyledons where photosynthate conduction is not so rapid, compound sieve plates tend to predominate. I have observed that pedicels of *Nuphar* (Nymphaeaceae) have sieve-tube elements with simple sieve plates, but only tracheids in the xylem. This corresponds to rapid input of photosynthates into the massive flower and fruit.

The above situation may hold in various dicotyledons. For example, I have been puzzled by curious non-conformities between vessel-element morphology and sieve-tube–element morphology, non-conformities that do not seem explainable on the basis of differential evolution rates, at least when framed in terms of progressive unidirectional trends. For example, *Fitchia* (Asteraceae) has, as do all Asteraceae, simple perforation plates, whereas secondary phloem clearly has long compound sieve plates, with some sieve areas partially subdivided by wall material. The same contrast between simple perforation plates in xylem and compound sieve plates in phloem obtains in Myrtaceae, Proteaceae, and Caricaceae (Zahur, 1959). These are all families which one would rate on wood and other characteristics as phylogenetically specialized, unlikely to retain primitive phloem configurations.

On the contrary, simple sieve plates occur in the same stems as vessels with long scalariform perforation plates in Cornaceae (*Cornus* mas), Actinidiaceae (*Saurauia* sp.) and Aquifoliaceae (Zahur, 1959). There is a possibility that simple sieve plates may occur in mesic species with long fusiform cambial initials, but these simple sieve plates are related to the septation of derivatives into a vertical series of sieve tube elements (Cercidiphyllaceae, Aquifoliaceae, etc.). This cuts down conductive efficiency, so that the occurrence of simple sieve plates in these compensates by restoring greater conductive potential.

The explanation for non-conformities between sieve plates and perforation plates within particular stems can be found in the tendency, in many stems, for a low rate of photosynthate conduction, so that there is no strong pressure for differentiation of the end wall into a simple sieve plate. Thus, the oblique overlap area corresponding to the overlap of the fusiform cambial initials from which the sieve tube elements were derived would tend to differentiate minimally.

Efficient simple sieve plates with large pores and large sieve-tube diameter would be expected in stems where a finite amount of conductive tissue must suffice for stems lacking secondary growth: *Cucurbita*, lianas and vines (both in monocotyledons and dicotyledons).

Just as rates of water flow in the xylem of conifers are typically slow, so probably are rates of photosynthate flow in conifer phloem. This would explain the minimal differentiation of end walls in conifer sieve cells, where end walls are much like lateral walls.

At any rate, the evidence available points to reversibility of the sieve plate. As I pointed out earlier (1961a), subdivision of a simple sieve plate into several sieve areas does not seem to involve processes like subdivision of a simple perforation plate into a scalariform one by the creation of bars. There is reason to believe that a species in which no scalariform perforation plates occur cannot give rise to a species with scalariform perforation plates, but the same does not hold with respect to simple and compound sieve plates. At any rate, future studies ought to take into account the potential correlation between photosynthate conduction characteristics and sieve-tube element morphology. The data of Cheadle and Whitford (1941) have such strong implications in this regard that we can no longer afford to overlook physiological correlates.

We can probably dismiss Zahur's (1959) conclusion that sieve-tube elements are more primitive in less specialized woody dicotyledons as a *prima facie* conclusion. This conclusion is probably an artifact based on the tendency for long sieve-tube elements to occur in woods with long fusiform cambial initials. The factors influencing length of fusiform cambial initials, as stated in earlier chapters, seem to depend primarily on physiologically and mechanically optimal lengths of tracheary elements in any particular species, and sieve-tube element length is probably a passive expression of those factors.

Translocation velocities in phloem tend to fall in the range of 10 to 100 cm. per hour, although rates as high as 300 cm. per hour have been measured in the soybean seedling (Zimmermann, 1960). Rates of translocation in xylem are much higher than in phloem. For example, Scholander, Hemmingsen and Garey (1961) reported a rate of 7,200 cm. per hour in the xylem of the rattan palm. Admitting that the rates of conduction in the xylem of a lianoid stem is much more rapid than that of typical trees, comparisons of figures cited by various authors for rates of flow in xylem and phloem show that, very likely, rates of flow in xylem are about 10 times more rapid (or on that order of magnitude) than rates of flow in phloem for any particular species. One can envision easily that even with much less rapid rates of flow in the phloem than xylem, some species will evolve simple sieve plates to accommodate volume and rapidity of conduction. However, the tension of water columns, as in xylem, with its manifold effects on xylem structural characteristics, has no counterpart in phloem.

MECHANICAL TISSUE OF PHLOEM AND CORTEX

In considering the mechanical characteristics of xylem, we would be mistaken if we overlooked the mechanical function provided by formation of fibers in protophloem, metaphloem, and secondary phloem of dicotyledons. Protophloem fibers often mature into thick-walled lignified cells as metaxylem tracheids and vessels are matured. In a primary stem, the mechanical strength provided by extraxylary fibers may well outweigh the mechanical strength of imperforate elements within the xylem. For example, *Linum* has long, intrusive fibers that are very thick walled and may account for a considerable portion of the strength of a stem (Esau, 1965). In the instance of an herb like *Linum*, production of mechanical tissue in phloem may well permit xylem to evolve chaiacteristics according to factors other than self-support of the plant, resistance to tension and shear. Thus, we could imagine that in a short-lived stem with considerable extraxylary fiber formation, short, relatively narrow vessel elements structurally ideal for counteracting negative pressures in xylem might be formed, and the mechanically strong fibers would occur not in the xylem, but in extraxylary regions.

This has several interesting implications. First, the less secondary growth a stem experiences, the more potentially significant extraxylary fibers and cortical sclerenchyma become. In woody plants, extraxylary fibers have little function in resistance to loading; and the more secondary xylem is present, the more extraxylary sclerenchyma relates to the functions of sclerenchyma in bark formation—functions which are outside the scope of the present study. In dicotyledons that lack extraxylary fibers entirely (e.g., lobelioids), imperforate elements in secondary xylem will, on the contrary, be a sensitive indicator of solutions to mechanical requisites of the growth form.

In monocotyledons, division of labor between xylem and extraxylary sclerenchyma is virtually complete. Bundle sheath fibers and ground tissue sclerenchyma compensate for the absence of mechanical tissue in xylem. Monocotyledons have a very flexible scheme, therefore, for conversion of bundle sheaths and ground tissues into amounts of sclerenchyma appropriate to any given growth form. Examples from palms (Tomlinson, 1961) are particularly appealing in this respect. The length of bundle-sheath fibers is much like that of tracheary elements in monocotyledons (see fig. 10, for example). However, unlike secondary xylem, there is no factor restricting the quantity of extraxylary sclerenchyma that may be produced. Monocotyledonous flexibility in response to mechanical function can also be seen in the distribution of sclerenchyma within an organ. For example, the majority of sclerenchyma in a bamboo stem is close to the periphery of a stem. In a mechanical root, such as that of an epiphytic orchid, mechanical tissues sheathe the stele, and are thus located well inward from the periphery of a root; in some monocotyledon roots, mechanical tissue in the pith is prominent (Carlquist, 1966b).

To be sure, a mechanical viewpoint on fibers in relation to plant architecture is anything but new. Notable contributions have been made by Schwendener (1874) and Rasdorsky (1929, 1937). For other references, the reader interested in developing a structural viewpoint of the plant in mechanical terms will wish to consult the literature cited by Tobler (1957). These essays remind us that a functional view of plant structure is not new. They also stand, however, as a challenge to plant anatomists, plant physiologists, and plant chemists to integrate new findings into a widening synthesis.

Literature Cited

Adams, J. E. 1949. Studies in the comparative anatomy of the Cornaceae. Jour. Elisha Mitchell Soc. 65: 218–244.

Airy-Shaw, H. K. 1941. The Pentaphragmataceae of the Oxford University expedition to Sarawak, 1932. Kew Bull. 3: 233–236.

Altman, P. L., and Dorothy S. Dittmer. 1966. Environmental Biology. Bethesda (Maryland), Federation of American Societies for Experimental Biology.

Anderson, L. C. 1972. Studies in *Bigelowia* (Asteraceae), II. Xylary comparisons, woodiness, and paedomorphosis. Jour. Arnold Arb. 53: 499–514.

Aung, M. 1962. Density variation outwards from the pith in some species of *Shorea* and its anatomical basis. Empire Forestry Review 41: 48–56.

Ayensu, E. S. 1972. Dioscoreales. Anatomy of the Monocotyledons, VI (C. R. Metcalfe, ed.). Oxford, Oxford University Press.

Ayensu, E. S., and Stern, W. L. 1964. Systematic anatomy and ontogeny of the stem in Passifloraceae. Contrib. U.S. Nat. Herb. 34(3): 45–73.

Bass, P. 1973. The wood anatomical range in *Ilex* (Aquifoliaceae) and its ecological and phylogenetic significance. Blumea 21: 193–258.

Bailey, I. W. 1944a. The comparative morphology of the Winteraceae. III. Wood. Jour. Arnold Arb. 25: 97–103.

Bailey, I. W. 1944b. The development of vessels in angiosperms and its significance in morphological research. Amer. Jour. Bot. 31: 421–428.

Bailey, I. W. 1953. Evolution of the tracheary tissue of land plants. Amer. Jour. Bot. 40: 4–8.

Bailey, I. W. 1957. The potentialities and limitations of wood anatomy in the study of phylogeny and classification of angiosperms. Jour. Arnold Arb. 38: 243–254.

Bailey, I. W. 1958. The structure of tracheids in relation to the movement of liquids, suspensions, and undissolved gases. In, K. V. Thimann, ed., Physiology of Forest Trees (pp. 71–82). New York, Ronald Press.

Bailey, I. W. 1966. The significance of the reduction of vessels in the Cactaceae. Jour. Arnold Arb. 47: 288–292.

Bailey, I. W., and Faull, A. F. 1934. The cambium and its derivative tissues. IX. Structural variability in the redwood, *Sequoia sempervirens*, and its significance in the identification of fossil woods. Jour. Arnold Arb. 15: 233–254.

Bailey, I. W., and Howard, R. A. 1941. The comparative morphology of the Icacinaceae. III. Imperforate tracheary elements and xylem parenchyma. Jour. Arnold Arb. 22: 432–422.

Bailey, I. W., and Nast, C. G. 1944. The comparative morphology of the Winteraceae. V. Foliar epidermis and sclerenchyma. Jour. Arnold Arb. 25: 342–348.

Bailey, I .W., and Shepard, H. B. 1915. Sanio's laws for the variation in size of coniferous tracheids. Bot. Gaz. 60: 66–71.

Bailey, I. W., and Swamy, B. G. L. 1948. *Amborella trichopoda* Baill., a new morphological type of vesselless dicotyledon. Jour. Arnold Arb. 29: 245–254.

Bailey, I. W., and Swamy, B. G. L. 1949. The morphology and relationships of *Austrobaileya*. Jour. Arnold Arb. 30: 211–226.

Bailey, I. W., and Tupper, W. W. 1918. Size variation in tracheary cells. I. A comparison between the secondary xylems of vascular cryptogams, gymnosperms and angiosperms. Proc. Amer. Acad. Arts & Sci. 54: 149–204.

Banks, C. C. 1973. The strength of trees. Jour. Inst. Wood Sci. 6(2): 44–50.

Bannan, M. W. 1941. Wood structure of *Thuja occidentalis*. Bot. Gaz. 103: 295–309.

Bannan, M. W. 1942. Wood structure of the native Ontario species of *Juniperus*. Amer. Jour. Bot. 29: 245–252.

Bannan, M. W. 1965. The length, tangential diameter and length/width ratio of conifer tracheids. Can. Jour. Bot. 43: 967–984.

Bannan, M. W. 1967. Sequential changes in rate of anticlinal division, cambial cell length and ring width in the growth of coniferous stems. Can. Jour. Bot. 54: 1359–1369.

Bannan, M. W. 1968. The problem of sampling in studies of tracheid length of conifers. Forest Sci. 14: 140–147.

Barghoorn, E. S. 1940. The ontogenetic and phylogenetic specialization of rays in the xylem of dicotyledons. I. The primitive ray structure. Amer. Jour. Bot. 27: 918–928.

Barghoorn, E. S. 1941a. The ontogenetic and phylogenetic specialization of rays in the xylem of dicotyledons. II. Modification of the multiseriate and uniseriate rays. Amer. Jour. Bot. 28: 273–282.

Barghoorn, E. S. 1941b. The ontogenetic and phylogenetic specialization of rays in the xylem of dicotyledons. III. The elimination of rays. Bull. Torrey Bot. Club 68: 317–325.

Barghoorn, E. S. 1964. Evolution of cambium in geologic time. In, M. H. Zimmermann, ed., The Formation of Wood in Forest Trees (pp. 3–17). New York and London, Academic Press.

Beijer, J. D. 1927. Die Vermehrung der radialen Reihen in Cambium. Rev. Trav. Bot. Néerland. 24: 631–786.

Bierhorst, D. W. 1958a. Vessels in *Equisetum*. Amer. Jour. Bot. 45: 534–537.

Bierhorst, D. W. 1958b. The tracheary elements of *Equisetum* with observations on the ontogeny of the internodal xylem. Bull. Torrey Bot. Club 85: 416–433.

Bierhorst, D. W. 1960. Observations on tracheary elements. Phytomorphology 10: 249–305.

Bierhorst, D. W. 1971. Morphology of Vascular Plants. New York, The Macmillan Co.

Bierhorst, D. W., and Zamora, P. M. 1965. Primary xylem elements and element associations of angiosperms. Amer. Jour. Bot. 52: 657–710.

Bongers, J. M. 1973. Epidermal leaf characters of the Winteraceae. Blumea 21: 381–411.

Bosshard, H. H. 1951. Variabilität der Elemente des Eschenholzes in Funktion von der Kambiumtätigkeit. Schweiz. Zeitschr. Forstw. 102: 648–665.

Boureau, E. 1964. Traité de Paléobotanique. II. Bryophyta, Psilophyta, Lycophyta. Paris, Masson & Cie.

Boureau, E. 1967. Traité de Paléobotanique. III. Sphenophyta, Noeggerathiophyta. Paris, Masson & Cie.

Boureau, E. 1970. Traité de Paléobotanique. IV. Filicophyta. Paris, Masson & Cie.

Bowen, G. D. 1973. Mineral nutrition of ectomycorrhizae. In,

G. C. Marks and T. T. Kozlowski, eds., Ectomycorrhizae (pp. 151–205). New York and London, Academic Press.

Bower, F. O. 1923. The Ferns (Filicales). Vol. I. Analytical Examination of the Criteria of Comparison. Cambridge, Cambridge University Press.

Bower, F. O. 1926. The Ferns (Filicales). Vol. II. The Eusporangiate and other Relatively Primitive Ferns. Cambridge, Cambridge University Press.

Bower, F. O. 1928. The Ferns (Filicales). Vol. III. The Leptosporangiate Ferns. Cambridge, Cambridge University Press.

Braun, H. J. 1970. Funktionelle Histologie der sekundären Sprossachse. I. Das Holz. Handbuch der Pflanzenanatomie IX(1): 1–190. Berlin and Stuttgart, Gebrüder Borntraeger.

Braun, H. J., and Wolkinger, F. 1970. Zur Funktionelle Anatomie des axialen Holzparenchyms und Vorschläge zur Reform seiner Terminologie. Holzforschung 24: 19–26.

Brown, F. B. H. 1922. The secondary xylem of Hawaiian trees. Occas. Pap. B. P. Bishop Mus. 7(6): 217–371.

Carlquist, S. 1957. Wood anatomy of Mutisieae (Compositae). Trop. Woods 106: 29–45.

Carlquist, S. 1958. Wood anatomy of Heliantheae (Compositae). Trop. Woods 108: 1–30.

Carlquist, S. 1961a. Comparative Plant Anatomy. New York, Holt, Rinehart & Winston.

Carlquist, S. 1961b. Wood anatomy of Inuleae (Compositae). Aliso 5: 21–37.

Carlquist, S. 1962. A theory of paedomorphosis in dicotyledonous woods. Phytomorphology 12: 30–45.

Carlquist, S. 1966a. Wood anatomy of Compositae: a summary, with comments on factors controlling wood evolution. Aliso 6(2): 25–44.

Carlquist, S. 1966b. Anatomy of Rapateaceae—roots and stems. Phytomorphology 16: 17–38.

Carlquist, S. 1969a. Wood anatomy of Goodeniaceae and the problem of insular woodiness. Ann. Missouri Bot. Gard. 56: 358–390.

Carlquist, S. 1969b. Wood anatomy of Lobelioideae (Campanulaceae). Biotropica 1: 47–72.

Carlquist, S. 1970a. Wood anatomy of *Echium* (Boraginaceae) Aliso 7: 183–199.

Carlquist, S. 1970b. Wood anatomy of Hawaiian, Macaronesian, and other species of *Euphorbia*. Bot. Jour. Linn. Soc. 63(Suppl. 1): 181–193.

Carlquist, S. 1970c. Wood anatomy of insular species of *Plantago* and the problem of raylessness. Bull. Torrey Bot. Club 97: 353–361.

Carlquist, S. 1971. Wood anatomy of Macaronesian and other Brassicaceae. Aliso 7: 365–384.

Carlquist, S. 1974. Island Biology. New York, Columbia University Press.

Caughey, M. G. 1945. Water relations of pocosin or bog shrubs. Plant Physiology 20: 671–689.

Cheadle, V. I. 1942. The occurrence and types of vessels in the various organs of the plant in the Monocotyledoneae. Amer. Jour. Bot. 29: 441–450.

Cheadle, V. I. 1943a. The origin and trends of specialization of the vessel in the Monocotyledoneae. Amer. Jour. Bot. 30: 11–17.

Cheadle, V. I. 1943b. Vessel specialization in the late metaxylem of the various organs in Monocotyledoneae. Amer. Jour. Bot. 30: 484–490.

Cheadle, V. I. 1944. Specialization of vessels within the xylem of each organ in the Monocotyledoneae. Amer. Jour. Bot. 31: 81–92.

Cheadle, V. I. 1953. Independent origin of vessels in the monocotyledons and dicotyledons. Phytomorphology 3: 23–44.

Cheadle, V. I. 1955. The taxonomic use of specialization of vessels in the metaxylem of Gramineae, Cyperaceae, Juncaceae, and Restionaceae. Jour. Arnold Arb. 36: 141–157.

Cheadle, V. I. 1963. Vessels in Iridaceae. Phytomorphology 13: 245–258.

Cheadle, V. I. 1968. Vessels in Haemodorales. Phytomorphology 18: 412–420.

Cheadle, V. I. 1969. Vessels in Amaryllidaceae and Tecophilaeaceae. Phytomorphology 19: 8–16.

Cheadle, V. I. 1970. Vessels in Pontederiaceae, Ruscaceae, Smil-

acaceae, and Trilliaceae. Bot. Jour. Linn. Soc. 63(Suppl. 1):
45–50.

Cheadle, V. I., and Kokasai, H. 1971. Vessels in Liliaceae. Phytomorphology 21: 320–333.

Cheadle, V. I., and Kokasai, H. 1972. Vessels in the Cyperaceae. Bot. Gaz. 133: 214–223.

Cheadle, V. I., and Whitford, N. B. 1941. Observations on the phloem in the Monocotyledoneae. I. The occurrence and phylogenetic specialization of the sieve tubes in the metaphloem. Amer. Jour. Bot. 623–627.

Coster, C. 1937. De verdamping van verschillende vegetativormen op Java. Tectona 30: 1–124.

Cumbie, B. G. 1960. Anatomical studies in the Leguminosae. Trop. Woods 113: 1–47.

Cumbie, G. B. 1963. The vascular cambium and xylem development in *Hibiscus lasiocarpus*. Amer. Jour. Bot. 50: 944–951.

Cumbie, B. G. 1967a. Developmental changes in the vascular cambium in *Leitneria floridana*. Amer. Jour. Bot. 54: 414–424.

Cumbie, B. G. 1967b. Development and structure of the xylem in *Canavalia* (Leguminosae). Bull. Torrey Bot. Club 94: 162–175.

Cumbie, B. G. 1969. Developmental changes in the vascular cambium of *Polygonum lapathifolium*. Amer. Jour. Bot. 56: 139–146.

Cumbie, B. G., and Mertz, D. 1962. Xylem anatomy of *Sophora* (Leguminosae) in relation to habit. Amer. Jour. Bot. 49: 33–40.

Cutler, D. F. 1969. Juncales. Anatomy of the Monocotyledons, IV, C. R. Metcalfe, ed. Oxford, Oxford University Press.

Dadswell, H. E., and H. D. Ingle. 1954. The wood anatomy of New Guinea *Nothofagus* Bl. Austral. Jour. Bot. 2: 141–153.

Dallimore, W., and Jackson, A. B. 1966. A Handbook of Coniferae and Ginkgoaceae, rev. by S. G. Harrison. London, Edward Arnold, Ltd.

Davis, T. A. 1961. High root-pressures in palms. Nature 192: 277–278.

De Zeeuw, C. 1965. Variability in wood. In, W. A. Côté, ed., Cellular Ultrastructure of Woody Plants (pp. 457–471). Syracuse (N.Y.), Syracuse University Press.

Dinwoodie, J. M. 1961. Tracheid and fiber length in timber, a review of literature. Forestry 34: 125–144.

Dinwoodie, J. M. 1963. Variation in tracheid length in *Picea sitchensis* Carr. Forest Prod. Spec. Rep. 16, D.S.I.R.

Dixon, H. 1914. Transpiration and the ascent of sap in plants. London, Macmillan and Co., Ltd.

Duerden, H. 1934. On the occurrence of vessels in *Selaginella*. Ann. Bot., ser. 1, 48: 459–465.

Epstein, E. 1972. Mineral Nutrition of Plants: Principles and Perspectives. New York, John Wiley & Sons, Inc.

Esau, K. 1965. Plant Anatomy. ed. 2. New York, John Wiley & Sons.

Fegel, A. C. 1941. Comparative anatomy and varying physical properties of trunk, branch, and root wood in certain northeastern trees. N.Y. State Coll. Forestry, Techn. Bull. 55: 1–20.

Florin, R. 1951. Evolution in cordaites and conifers. Acta Horti Bergiani 15: 285–388.

Florin, R. 1963. The distribution of conifer and taxad genera in time and space. Acta Horti Bergiani 20: 121–312.

Frei, E. 1955. Die Innervierung des floralen Nektarien dikotylen Pflanzenfamilien. Ber. Schweiz. Bot. Ges. 65: 60–114.

Frost, F. H. 1930a. Specialization in secondary xylem in dicotyledons. I. Origin of vessel. Bot. Gaz. 89: 67–94.

Frost, F. H. 1930b. Specialization in secondary xylem in dicotyledons. II. Evolution of end wall of vessel segment. Bot. Gaz. 90: 198–212.

Frost, F. H. 1931. Specialization in secondary xylem in dicotyledons. III. Specialization of lateral wall of vessel segment. Bot. Gaz. 91: 88–96.

Gates, D. M. 1968. Transpiration and leaf temperature. Ann. Rev. Plant Physio. 19: 211–238.

Gibson, A. C. 1973. Wood anatomy of Cactoideae (Cactaceae). Biotropica 5: 29–65.

Gilbert, S. G. 1940. Evolutionary significance of ring porosity in woody angiosperms. Bot. Gaz. 102: 105–120.

Greenidge, K. N. H. 1957. Ascent of sap. Ann. Rev. Plant Physio. 8: 237–256.

Greguss, P. 1955. Identification of Living Gymnosperms on the Basis of Xylotomy. Budapest, Akadémiai Kiadó.

Greguss, P. 1959. Holzanatomie der Laubhölzer und Sträucher. Budapest, Akadémiai Kiadó.

Greguss, P. 1968. Xylotomy of the Living Cycads. Budapest, Akadémiai Kiadó.

Groom, P. 1910. Remarks on oecology of Coniferae. Ann. Bot., ser. 1, 24: 241–269.

Hale, J. D., and Clermont, L. P. 1963. Influence of parenchyma cell-wall morphology on basic physical and chemical characteristics of wood. Jour. Polymer Sci., part C, 2: 253–261.

Hata, K. 1949. Studies on the pulp of akamatsu (Pinus densiflora Sieb. & Zucc.). I. On the length, diameter and length-diameter ratio of tracheids in the akamatsu wood. Kanagawa-ken Agr. Coll. Techn. Bull. 1: 1–35.

Henes, E. 1959. Fossile Wandstruktur. Handbuch der Pflanzenanatomie III(5): 1–108. Berlin-Nikolassee, Gebrüder Borntraeger.

Henrickson, J. S. 1968. Vegetative morphology of the Fouquieriaceae. Thesis, Claremont Graduate School.

Hirmer, M. 1927. Handbuch der Paläobotanik. I. Thallophyta—Bryophyta—Pteridophyta. München & Berlin, R. Oldenbourg.

Huber, B. 1935. Die physiologische Bedeutung der Ring und Zerstreutporigkeit. Ber. Deutsch Bot. Ges. 53: 711–719.

Huber, B. 1953. Was wissen wir vom Wasserverbrauch des Waldes. Forstw. Centr. 72: 257–264.

Huber, B. 1956. Die Gefässleitung. In, W. Ruhland, ed., Handbuch der Pflanzenphysiologie 3: 541–582. Berlin, Springer Verlag.

Ingle, H. D., and Dadswell, H. E. 1953. The anatomy of the timbers of the southwest Pacific area. II. Apocynaceae and Annonaceae. Austral. Jour. Bot. 1: 1–26.

Jane, F. W. 1956. The Structure of Wood. New York, The Macmillan Co.

Janssonius, H. H. 1906–1936. Mikrographie des Holzes der auf Java Vorkommenden Baumarten. 6 vols. Leiden, E. J. Brill.

Jeffree, C. E., Johnson, R. P. C., and Jarvis, P. G. 1971. Epicuticular wax in the stomatal antechamber of Sitka spruce and its effects on the diffusion of water vapor and carbon dioxide. Planta 98: 1–10.

Kanehira, R. 1921a. Anatomical Characters and Identification of Formosan Woods with Critical Remarks from the Climatic Point of View. Taihoku (Taiwan), Bureau of Productive Industries.

Kanehira, R. 1921b. Identification of the Important Japanese Woods by Anatomical Characters. Supplement to the Anatomical Characters and Identification of Formosan Woods. Taihoku (Taiwan), Bureau of Productive Industries.

Kanehira, R. 1924. Identification of Philippine Woods by Anatomical Characters. Taipei (Taiwan), Gov't. Research Institute of Formosa.

Kokasai, H., M. F. Moseley, Jr., and V. I. Cheadle. 1970. Morphological studies of the Nymphaeaceae. V. Does Nelumbo have vessels? Amer. Jour. Bot. 57: 487–494.

Kozlowski, T. T. 1943. Transpiration of some forest trees during the dormant season. Plant Physio. 18: 252–260.

Kramer, P. J. 1952. Plant and soil water relations on the watershed. Jour. Forestry 50: 92–95.

Kramer, P. J. 1959. Tarnspiration and the water economy of plants. In, F. C. Steward, ed., Plant Physiology, Vol. II. (pp. 607–730). New York, Academic Press.

Kribs, D. A. 1935. Salient lines of structural specialization in the wood rays of dicotyledons. Bot. Gaz. 96: 547–557.

Kribs, D. A. 1937. Salient lines of structural specialization in the wood parenchyma of dicotyledons. Bull. Torrey Bot. Club 64: 177–186.

Larson, P. R. 1962. The indirect effect of photoperiod on tracheid diameter in Pinus resinosa. Amer. Jour. Bot. 49: 132–136.

Läuchli, A. 1972. Translocation of organic solutes. Ann. Rev. Plant Physio. 23: 197–218.

Lee, C. L. 1961. Crystallinity of wood cellulose fibers studied by X-ray methods. Forest Prod. Jour. 11: 8–12.

Lemoigne, Y. 1961. Études analytiques et comparées des structures internes des Sigillaires. Thesis, Faculté des Sciences, Lille.

Liese, W. 1965. The fine structure of bordered pits in softwoods. In, W. A. Côté, ed., Cellular Ultrastructure of Woody

Plants (pp. 271–290). Syracuse (N.Y.), Syracuse University Press.

Lutz, H. J. 1952. Occurrence of clefts in the wood of living white spruce in Alaska. Jour. Forestry 50: 99–102.

MacDougal, D. T. 1921. Growth in Trees. Carnegie Institute of Washington Publ. 307. Washington, D.C., Carnegie Institute of Washington.

Mark, R. 1965. Tensile stress analysis of the cell walls of coniferous tracheids. In, W. A. Côté, ed., Cellular Ultrastructure of Woody Plants (pp. 493–533). Syracuse (N.Y.), Syracuse University Press.

Marks, G. C., and T. T. Kozlowski, eds. 1973. Ectomycorrhizae. New York and London, Academic Press.

Mell, C. D. 1910. Determination of quality locality by fiber length of wood. Forest. Quart. 8: 419–422.

Metcalfe, C. R. 1971. Cyperaceae. Anatomy of the Monocotyledons, V, C. R. Metcalfe, ed. Oxford, Oxford University Press.

Metcalfe, C. R., and Chalk, L. 1950. Anatomy of the Dicotyledons. 2 vols. Oxford, Clarendon Press.

Meyer, F. H. 1973. Distribution of ectomycorrhizae in native and man-made forests. In, G. C. Marks and T. T. Kozlowski, eds., Ectomycorrhizae (pp. 79–105). New York and London, Academic Press.

Neales, T. F., Patterson, A. A., and Hartney, V. J. 1968. Physiological adaptation to drought in the carbon assimilation and water loss of xerophytes. Nature 219: 469–472.

Novruzova, Z. A. 1968. The Water Conducting System of Trees and Shrubs in Relation to Ecology. Baku (U.S.S.R.), Izdatel'stvo Akademii Nauk Azerbaijan S.S.R.

Ogura, Y. 1972. Comparative Anatomy of Vegetative Organs of the Pteridophytes. Handbuch der Pflanzenanatomie VII(3): 1–502. Berlin and Stuttgart, Gebrüder Borntraeger.

Parker, J. 1956. Drought resistance in woody plants. Bot. Rev. 22: 241–289.

Patel, R. N. 1965. A comparison of the anatomy of the secondary xylem in roots and stems. Holzforschung 19: 72–79.

Patel, R. N. 1973. Wood anatomy of the dicotyledons indigenous to New Zealand. I. Cornaceae. New Zealand Jour. Bot. 11: 3–22.

Peel, A. J. 1965. On the conductivity of the xylem in trees. Ann. Bot., ser. 2, 29: 119–130.

Phillips, E. W. J. 1941. The identification of coniferous woods by their microscopic structure. Jour. Linn. Soc. London 52: 259–320.

Rasdorsky, W. 1929. Über die Baumechanik der Pflanzen. Biologia Generalis 1: 63–94.

Rasdorsky, W. 1937. Über die Baumechanik der Pflanzen. Biologia Generalis 12: 359–398.

Ray, P. M. 1972. The Living Plant. ed. 2. New York, Holt, Rinehart & Winston, Inc.

Record, S. J. 1919. Identification of the Economic Woods of the United States. New York, John Wiley & Co.

Renault, B. 1885. Recherches sur les végétaux fossiles du genre *Astromyelon*. Ann. Soc. Géol. Paris 17: 1–34.

Renault, B. 1898. Notice sur les Calamariacées. Bull. Soc. Hist. Nat. Autun 11: 377–436.

Richardson, D. D. 1964. The external environment and tracheid size in conifers. In, M. H. Zimmermann, ed., The Formation of Wood in Forest Trees (pp. 367–388). New York and London, Academic Press.

Ritter, G. J., and Fleck, L. C. 1926. Chemistry of wood. IX. Springwood and summerwood. Indus. Eng. Chem. 18: 608–609.

Rivett, M. F. 1920. The anatomy of *Rhododendron ponticum* L. and of *Ilex aquifolium* L. in reference to specific conductivity. Ann. Bot., ser. 1, 34: 525–550.

Rumball, W. 1963. Wood structure in relation to heteroblastism. Phytomorphology 13: 206–214.

Salisbury, F. B., and Ross, C. 1969. Plant Physiology. Belmont (Calif.), Wadsworth Publishing Co., Inc.

Sanio, K. 1872. Über die Grösse der Holzzellen bei der gemeinen Kiefer (*Pinus silvestris*). Jahrb. Wiss. Bot. 8: 401–420.

Sastrapadja, D. S., and Lamoureux, C. 1969. Variations in wood anatomy of Hawaiian *Metrosideros* (Myrtaceae). Ann. Bogorienses 5(1): 1–83.

Scholander, P. F., Hammel, H. T., Bradstreet, E. D., and Hem-

240   LITERATURE CITED

mingsen, E. A. 1965. Sap pressure in vascular plants. Science 148: 339–345.

Scholander, P. F., Hammel, H. T., Hemmingsen, E. A., and Garey, W. 1962. Salt balance in mangroves. Plant Physio. 37: 722–729.

Scholander, P. F., Hemmingsen, E. A., and Garey, W. 1961. Cohesive lift of sap in the rattan vine. Science 134: 1835–1838.

Scholander, P. F., Love, W. E., and Kenwisher, J. W. 1955. The rise of sap in tall grapevines. Plant Physio. 30: 93–104.

Scholander, P. F., Ruud, B., and Leivestad, H. 1957. The rise of sap in a tropical liana. Plant Physio. 32: 1–6.

Schweitzer, E. M. 1971. Comparative anatomy of Ulmaceae. Jour. Arnold Arb. 52: 523–585.

Schwendener, S. 1874. Das mechanische Prinzip in anatomische Bau der Monokotylen mit vergleichenden Ausblicken auf die übrigen Pflanzenklassen. Leipzig.

Siau, J. F. 1971. Flow in Wood. Syracuse (N.Y.), Syracuse University Press.

Smith, A. C. 1943a. The American species of Drimys. Jour. Arnold Arb. 24: 1–33.

Smith, A. C. 1943b. Taxonomic notes on the Old World species of Winteraceae. Jour. Arnold Arb. 24: 119–164.

Smith, A. C. 1945. A taxonomic review of Trochodendron and Tetracentron. Jour. Arnold Arb. 26: 123–142.

Soderstrom, T. R., and Calderon, C. E. 1971. Insect pollination in tropical rain forest grasses. Biotropica 3: 1–16.

Spurr, S. H., and Hyvärinen, M. J. 1954. Wood fiber length as related to position in tree and growth. Bot. Rev. 20: 561–575.

Starr, A. M. 1912. Comparative anatomy of dune plants. Bot. Gaz. 54: 265–305.

Steenis, C. G. G. J. van. 1963. Pacific Plant Areas. Vol. 1. Monographs of the National Institutes of Science and Technology, Manila, 8(1): 1–297.

Steenis, C. G. G. J. van, and Balgooy, M. M. J. van. 1966. Pacific Plant Areas. Vol. 2. Blumea, Suppl. Vol. 5: 1–312.

Stern, W. L., and Greene, S. 1958. Some aspects of variation in wood. Trop. Woods 108: 65–71.

Swamy, B. G. L. 1953. The morphology and relationships of the Chloranthaceae. Jour. Arnold Arb. 34: 375–408.

Swamy, B. G. L., and Bailey, I. W. 1950. *Sarcandra*, a vesselless genus of the Chloranthaceae. Jour. Arnold Arb. 31: 117–129.

Swamy, B. G. L., and Govindarajalu, E. 1957. Anatomical studies in polyploid strains of *Parthenium argentatum*. Jour. Asiatic Soc., Sci., 23(3): 43–54.

Swamy, B. G. L., and Govindarajalu, E. 1961. Studies on the anatomical variability in the stem of *Phoenix sylvestris*. I. Trends in the behavior of certain cells and tissues. Jour. Indian Bot. Soc. 40: 243–262.

Swamy, B. G. L., Parameswaran, N., and Govindarajalu, E. 1960. Variation in vessel length within one growth ring of certain arborescent dicotyledons. Jour. Indian Bot. Soc. 39: 163–170.

Tabata, H. 1964. Vessel element of Japanese birches as viewed from ecology and evolution. Physio. & Ecol. (Japan) 12: 7–16.

Takhtajan, A. 1969. Flowering Plants. Origin and Dispersal (C. Jeffrey, transl.). Edinburgh, Oliver & Boyd.

Tinklin, R. and Weatherley, P. E. 1968. The effect of transpiration rate on the leaf water potential of sand and soil-rooted plants. New Phytol. 67: 605–615.

Tobler, F. 1957. Die mechanische Elemente und das mechanische System. Handbuch der Pflanzenanatomie IV(6): 1–60. Berlin-Nikolassee, Gebrüder Borntraeger.

Tomlinson, P. B. 1961. Palmae. Anatomy of the Monocotyledons, II, C. R. Metcalfe, ed. Oxford, Oxford University Press.

Tomlinson, P. B. 1969. Commelinales–Zingiberales. Anatomy of the Monocotyledons, III, C. R. Metcalfe, ed. Oxford, Oxford University Press.

Tomlinson, P. B., and E. S. Ayensu. 1969. Notes on the vegetative morphology and anatomy of the Petermanniaceae (Monocotyledons). Bot. Jour. Linn. Soc. 62: 17–26.

Tomlinson, P. B., and Zimmermann, M. H. 1967. The "wood" of monocotyledons. Bull. Intern. Assoc. Wood Anat. 1967/2: 4–24.

Tomlinson, P. B., and Zimmerman, M. H. 1969. Vascular anatomy of monocotyledons with secondary growth—an introduction. Jour. Arnold Arb. 50: 159–179.

Tryon, R. M. 1955. *Selaginella rupestris* and its allies. Ann. Missouri Bot. Gard. 42: 1–99.

Tsoumis, G. 1965. Light and electron microscopic evidence on the structure of the membrane of bordered pits in tracheids of conifers. In, W. A. Côté, ed., Cellular Ultrastructure of Woody Plants (pp. 305–317). Syracuse (N.Y.), Syracuse University Press.

Van Teighem, P. 1886. Sur la polystélie. Ann. Sci. Nat. Bot., ser. 7, 3: 275–322.

Versteegh, C. 1968. An anatomical study of some woody plants of the mountain flora in the tropics (Indonesia). Acta Bot. Néerland. 17: 151–159.

Walsh, M. A. 1974. Xylem anatomy of *Hibiscus* (Malvaceae) in relation to habit. Can. Jour. Bot. (in press).

Walter, H., and Lieth, H. 1967. Klimadiagramm-Weltatlas. Jena, Gustav Fischer Verlag.

Wardrop, A. B. 1951. Cell wall organization and the properties of xylem. I. Cell wall organization and the variation of breaking load in tension of xylem in conifer stems. Austral. Jour. Sci. Res., ser. B, 4: 391–414.

Webber, I. E. 1936. The woods of sclerophyllous and desert shrubs and desert plants of California. Amer. Jour. Bot. 23: 181–188.

Wellwood, R. W. 1962. Tensile testing of small wood samples. Pulp Paper Mag. Can. 63(2):T61–T67.

White, R. A. 1961. Vessels in roots of *Marsilea*. Science 103: 1073–1074.

White, R. A. 1962. A comparative study of the tracheary elements of the ferns. Thesis, University of Michigan.

White, R. A. 1963a. Tracheary elements of the ferns. I. Factors which influence tracheid length; correlation of length with evolutionary divergence. Amer. Jour. Bot. 50: 447–456.

White, R. A. 1963b. Tracheary elements of the ferns. II. Morphology of tracheary elements; conclusions. Amer. Jour. Bot. 50: 514–522.

Williamson, W. C., and Scott, D. H. 1894. Further observations on the organization of the fossil plants of Coal-Measures. I. *Calamites*, *Calamostachys*, and *Sphenophyllum*. Phil. Trans. Roy. Soc. London, B, 185: 863–959.

Wilson, T. K. 1960. The comparative morphology of the Canellaceae. I. Synopsis of genera and wood anatomy. Trop. Woods 112: 1–27.

Winstead, J. E. 1972. Fiber tracheid length and wood specific gravity of seedlings as ecotypic characters in *Liquidambar styraciflua* L. Ecology 53: 165–172.

Wolkinger, F. 1969. Morphologie und systematische Verbreitung der lebenden Holzfasern bei Sträuchern und Bäumen. I. Zur Morphologie und Zytologie. Holzforschung 23: 135–144.

Wolkinger, F. 1970. Morphologie und systematische Verbreitung der lebenden Holzfasern bei Sträuchern und Bäumen. II. Zur Histologie. Holzforschung 24: 141–151.

Wolkinger, F. 1971. Morphologie und systematische Verbreitung der lebenden Holzfasern bei Sträuchern und Bäumen. Holzforschung 25: 29–30.

Wray, F. J., and Richardson, J. A. 1964. Paths of water transport in higher plants. Nature 202: 415–416.

Wulff, T. 1898. Studien über verstopfte Spaltöffnungen. Öst. Bot. Zeitschr. 48: 201–209, 252–258, 298–307.

Zahur, M. S. 1959. Comparative study of secondary phloem of 423 species of woody dicotyledons belonging to 85 families. Cornell Univ. Agr. Exp. Sta. Mem. 358: 1–160.

Zimmermann, M. H. 1960. Transport in the phloem. Ann. Rev. Plant Physiol. 11: 167–190.

Zimmermann, M. H. 1963. How sap moves in trees. Sci. Amer. 208(3): 132–142.

Zimmermann, M. H. 1964. Sap movements in trees. Biorheology 2: 15–27.

Zimmermann, M. H. 1965. Water movement in stems of tall plants. In, The State and Movement of Water in Living Organisms, Symposia of the Society for Experimental Biology 19(pp. 151–155). New York, Academic Press.

Zimmermann, M. H., and Brown, C. L. 1971. Trees. Structure and Function. New York, Springer Verlag.

Zimmermann, M. H., and Tomlinson, P. B. 1965. Anatomy of the palm *Rhapis excelsa*. I. Mature vegetative axis. Jour. Arnold Arb. 46: 160–178.

Index

Abronia (Nyctaginaceae), wood of, 212

Actinostrobus (Cupressaceae), shortness of tracheids of, 87

Adenophorus haalilioanus (Grammitidaceae), tracheids of, 34

Aeonium arboreum (Crassulaceae): raylessness in, 189; paedomorphosis in, 222

Aextoxicon punctatum (Aextoxicaceae): vessel data, 142–143; length ratios of tracheids to vessel elements, 167; age-on-length curve of tracheids, 218–219.

Agapanthus (Liliaceae), perforation plates scalariform in roots of, 109

Agathis (Araucariaceae): exceptional tracheid length of, 87; wide tracheids of, 91, 93; axial parenchyma abundant in, 99

Agave (Liliaceae), seasonality in, 29

Air embolisms. See Cavitations

Alismataceae, vessels specialized in roots of, 114

Alnus incana (Betulaceae), wood varies according to habitat of, 9

Aloë (Liliaceae), seasonality in, 29

Alpine conifers, wood of, 94

Alpine shrubs, wood of, 182, 208

Alsophila (Cyatheaceae), tracheids of, 35

Altitude in relation to wood anatomy, 10

"Alveolar occluding material." See Stomata, plugged

Amborella trichopoda (Amborellaceae): pits in ray cells, 14; nature and significance of tracheids in, 131–132, 136; rays of, 200

Anacardiaceae, woods in lowlands versus uplands, 10

Angular nature of vessels, 163–164

Annuals: dicotyledonous, outermost

wood xeromorphic in, 182; characteristics of wood of, 209

Anomalous secondary growth. See Successive cambia

Apocynaceae, woods in lowlands versus uplands, 10

Aponogetonaceae, data on xylem of, 29

Aquatic monocotyledons, 120–121

Araliaceae, Hawaiian, 147

Araucaria (Araucariaceae), tracheid diameter of, 91

Araucariaceae, height and tracheid length in, 92–93

Archaeopteris, tracheids in, 50

Arecaceae. See Palms

Artemisia arbuscula (Asteraceae): vessel characteristics of, 177; wood parenchyma of, 198

Arthropitys (Calamitales), 53–54

Aspidiaceae, tracheids in, 35, 41

Asteraceae (= Compositae): woods vary with ecology, 1; in wet Hawaiian forest, 147; secondary adaptations to mesic habitats, 180; helices in vessels of, 195; rays of, 203; woods compared on basis of habitat, 176

Aster spinosus (Asteraceae), exemplary of herbaceous wood, 217–219

Athrotaxis (Taxodiaceae), short tracheids of, 87

Athyrium niponicum (Aspidiaceae), dictyostele in, 73

Aucuba japonica (Cornaceae), starch in tracheids of, 197

Austrobaileya scandens (Austrobaileyaceae): nodes and leaf traces of, 77; wood specialization of accelerated by lianoid habit, 155; perforation plates in, 160, 161; accelerated specialization of vessel-

primitive and specialized dicotyledon woods in, 149–150
Specialization: trends of in wood of
dicotyledons, 12–22; trends of in
monocotyledons, 22–23; functional
explanation of in monocotyledons,
104–129; functional explanation of
in dicotyledons, 174–222
Sphenophyllum (Sphenophyllales),
51–53
Sphenopsida. *See* Calamitales; *Equisetum*; *Sphenophyllum*
Sphenostemon (Sphenostemonaceae):
vessel data on, 142, 143, 168;
length ratio of tracheids to vessel
elements, 166
Spirals (in secondary xylem tracheary
elements). *See* Helices
Starch. *See* Axial parenchyma; Photosynthate translocation; Rays
Statistical correlations, use of in evolutionary hypotheses, 1
Stelar theory, 61–78
Stems: wood of, compared with that
of roots in dicotyledons, 179. *See
also* Organ or portion of a plant;
Organographic differences in tracheary elements; Origin of vessels
Stenochlaena tenuifolia (Blechnaceae), dictyostele in, 73
Stomata, plugged: in conifers, 95–96;
in vesselless dicotyledons, 133–134
Storage organs, stelar modifications
in, 77–78
Storied wood, 190–192
Strength. *See* Mechanical strength
Stromatopteris. See Gleicheniaceae
Stylites. See Isoëtes
Successive cambia: in storage organs,
78; conductive characteristics of
woods with, 211
Succulents: *Isoëtes* compared to, 48;
cycads a type of, 79, 81; relationship to xylem in monocotyledons,
110–111; nocturnal transpiration
in, 112; compared to *Pentaphragma*, 157; low tensions in water

columns of, 187–188; shrinkability
of tracheary elements in, 188; significance of large rays in, 203; conductive characteristics in wood of,
207–208; release of mechanical
strength in wood of, 217–219. *See
also* Cactaceae
Swamps. *See* Marshes
Symplocos (Symplocaceae), vessel
data in, 142–143, 153

Tabular comparisons, use of in com-
parative study of woods, 9
Tacca (Taccaceae), vessels of indicative of mesomorphy, 114
Taiwan, nature of woods in, 146
Talinum guadalupense (Portulaca-
ceae), release of mechanical strength
in, 217–219
Tasmannia. See Winteraceae
Taxaceae: spirals in, 102. *See also*
Conifers
Taxodiaceae: relict nature of, 98; non-
mycorrhizal nature of, 103. *See
also* under particular genera
Taxodium mucronatum (Taxodiaceae), wood characteristics and
ecology, 90, 93
Tensions in water columns of xylem:
in protosteles, 63; in conifer-root
tracheids, 88; notably high, 92;
probable distribution of in palm
stem, 125–126; in desert shrubs,
176; high in *Sequoia*, 177; lower
in roots than in stems, 179; low in
lianas, 181; low in succulents, 187–
188
"Tertiary helical thickenings." *See*
Helices
Tetracentron sinense (Tetracentraceae): ecology compared to that of
Ginkgo, 82; nature and significance
of tracheids in, 131, 133, 134–137
Theaceae, vessel data on, 142–143,
153
Thuja occidentalis (Cupressaceae), in
swampy versus sandy soils, 8